Consumer Math

Harold Parks
Gary Musser
Robert Burton
William Siebler

Pearson Custom Publishing

Taken from:

Mathematics in Life, Society & the World, Second Edition,
by Harold Parks, Gary Musser, Robert Burton and William Siebler

A Pearson Education Company
Upper Saddle River, New Jersey 07458

This special edition published in cooperation with
Pearson Custom Publishing.

Printed in the United States of America

10 9 8 7 6 5 4 3 2 1

Please visit our web site at www.pearsoncustom.com

ISBN 0-536-61647-7

BA 992557

PEARSON CUSTOM PUBLISHING
75 Arlington Street, Boston, MA 02116
A Pearson Education Company

Contents

6

Consumer Mathematics—Buying and Saving

Analyst Predicts Stock Market Perfectly

Suppose you are considering investing in the stock market. However, you have not developed an investment strategy, nor have you selected a financial adviser. You receive a letter from a stock market consultant, J. J. Herringbone, touting his services. According to the letter, he has astounded the experts with his ability to predict the volatile commodities market. This is a market in which 90% of the investors lose money, yet he has successfully predicted the movements of prices over 80% of the time! As an inducement to subscribe to his services, he provides you with six predictions regarding the market for the next ten days. When the ten days are up, you are surprised to see that Herringbone has been correct on all of his predictions!

The next week, you receive another letter from Herringbone with an offer of his services. Now that the accuracy of his system has been verified, he is offering yearly memberships that will include a newsletter with future predictions. These memberships will cost $5000 per year. The letter ends with a postscript: "Individuals who have followed my earlier tips have already recovered the full cost of their membership and profited substantially. This is a limited time offer! Do not miss out on this once in a lifetime opportunity."

CHAPTER GOALS

1. *Calculate simple and compound interest.*
2. *Become familiar with various types of loans and how to compute finance charges (interest).*
3. *Learn about financing a house.*
4. *Find the future value of annuities.*

Does this sound like a good offer? If this is a fraud, how was Herringbone able to predict the market 6 times with such accuracy? The probability of doing this by chance is 1 in 64, which makes this very unlikely. Should you run out with the money you have been saving for a down payment on a house? No, you should not.

If something sounds too good to be true, then it probably is. Herringbone could have chosen 64 different cities in which to promote his services and newsletter. For his first prediction, half of the cities are sent the information that stock A will rise, and the other half got the news that stock A will fall. He will be correct in 32 cities. Herringbone could send 16 of these cities the prediction that stock B will rise and the other 16 the prediction that the stock will fall. He will be correct in 8 cities. He continues this process until he is left with one city—yours—in which he has made only correct predictions. Based only on the movement of the stock (rise or fall), Herringbone can develop 64 sets of predictions, one of which is 100% correct, and another six sets of predictions that are correct in five out of six predictions (83%).

This offer is deceptive and may be a crime. The world of stocks, bonds, and other investments is very complex. It is replete with offers suggesting methods you can use to get rich quickly with "little risk," which are just as strange as Herringbone's scam, but not necessarily illegal. The purpose of this chapter is to teach you how to use your money carefully, although it will not discuss speculative investments. The chapter shows how interest works, how finance options and strategies can help you, and the advantages and dangers in certain types of borrowing.

THE HUMAN SIDE OF MATHEMATICS

Charles Dow

The secret of acquiring great wealth on the stock market is well known: Buy low and sell high. The trick is knowing what the market will do. Charles Dow (1851–1902) was a financial journalist who believed that the stock market could be understood using mathematical principles. Dow was a founder of Dow Jones & Company, which became a financial empire. The company published an index averaging various stocks as an indicator of how the market is doing as a whole. This publication became the precursor of *The Wall Street Journal.* Dow invented the "Dow Theory" of buying stocks, which is based entirely on numerical market information. Dow theorists are active in the market to this day. Dow was a man of great confidence and energy. As a young man he applied for a reporter's job at the *Providence Journal* and was told there was nothing for him to do. He replied that this was all right—he knew what news was and would find things to do. He became their star reporter for the next five years.

Ralph Elliott

Another person who tried to understand the stock market using mathematical principles was Ralph Elliott (1871–1948). Elliott was an engineer by profession. In the 1930s, he studied the Great Pyramid at Giza and concluded that the design for the structure was based on the Fibonacci numbers, which include 2, 3, 5, 8, 13, 21, 34. He decided that this progression of numbers held the secret to predicting the ups and downs of the stock market. In 1927, Elliott contracted a severe illness and spent many years in convalescence. It was during this time that he developed a theory on price movements now known as the *Elliott Wave Theory.* Before you run out and purchase the secrets of Dow and Elliott, you should know that professional practitioners of these methods often come up with different predictions based on the same information. Today, there are thousands of people (called chartists or technicians) with advanced degrees trying to forecast the market using mathematics. In the high-powered world of the stock market, even the slightest advantage can often translate to instant riches.

6.1 INTEREST

INITIAL PROBLEM

You discover that you are the direct descendant of a man who loaned the Continental Congress $1000 in 1777. However, he was never repaid nor have any of his descendants received repayment. You think it is about time to get the family money back. How much should you demand from the U.S. government? Use an interest rate of 6% and a compounding period of 3 months.

History

In the past, charging interest on borrowed money was often considered evil, and, in particular, was long prohibited by the Catholic Church. One way people got around the law against paying interest was to borrow in one currency and repay in another, the interest being disguised in the exchange rate.

Not having the money you want when you want it is a too common experience for most people, whether it's having enough money for tuition, buying essentials for daily living, or buying a new car. In this chapter we look at the different ways in which money can be borrowed (or saved), beginning with the familiar concept of simple interest and progressing to fairly complex financial instruments called annuities.

Simple Interest

Many people deposit savings in a bank or similar financial institution; likewise, many borrow money for major purchases such as a house or a car. Credit cards are used routinely to borrow smaller amounts of money to make purchases. We expect to receive interest income on our savings, and we also expect to pay interest on the money we borrow. Thus it is important to understand how interest is calculated, so that you can make informed decisions.

Not surprisingly, the simplest type of interest calculation is called **simple interest**. To calculate simple interest, you must know the amount of money on which the interest is being charged, the interest rate, and the time period over which the interest is being charged. Suppose you borrow $1000 at 5% simple interest for two years. The amount of money on which the interest will be charged is called the **principal**, in this case $1000. The **interest rate**, expressed as a percent, is the percentage of the principal that will be paid each year, in this case 5%. A unit of time other than years could be specified, but "per year" is understood if nothing else is said. Since the interest is percent per year, one must know the **time period** in years over which the interest will be charged, in your case two years. The formula for calculating interest is given next.

SIMPLE INTEREST FORMULA

If I represents interest, P the principal, r the interest rate expressed as a decimal, and t the time, then

$$I = Prt.$$

In words,

$$\text{interest} = \text{principal} \times \text{rate} \times \text{time (in years)}.$$

The (interest) rate in this formula is the annual rate and is normally given as a percent. For the example of borrowing \$1000 at 5% simple interest for two years, you calculate the interest to be

$$\text{interest} = \$1000 \times 0.05 \times 2 = \$100.$$

Note: $r\% = (0.01)r$.

EXAMPLE 6.1 Find the interest on a loan of \$100 at 6% simple interest for 1, 2, and $2\frac{1}{2}$ years.

Solution

For one year the interest is $\$100 \times 0.06 \times 1 = \$6$. The interest for two years is $\$100 \times 0.06 \times 2 = \$12$. For each year the interest is 6% of the principal of \$100, that is, \$6 per year. Thus for $2\frac{1}{2}$ years the interest would be $\$6 \times 2\frac{1}{2} = \$15$.

Still assuming you have borrowed \$1000 at 5% simple interest for two years, at the end of the two-year loan period you will need to repay the \$1000 that you borrowed plus you will need to pay the lender the \$100 in interest that we just calculated. Thus the total you will need to pay to the lender is $\$1000 + \$100 = \$1100$. That \$1100 is called the **future value** of the loan. For a simple interest loan, such as the one being discussed, the future value is simply the sum of the principal and the interest. The future value is important, because it represents the amount of money that must be given to the lender at the end of the loan. The formula for calculating the future value is given next.

FUTURE VALUE FOR SIMPLE INTEREST

If F represents the future value, P the principal, I the interest, r the interest rate expressed as a decimal, and t the time, then

$$F = P + I = P + Prt = P(1 + rt).$$

EXAMPLE 6.2 Find the future value of a loan of \$400 at 7% simple interest for three years.

Solution

We use the formula for future value to compute

$$F = \$400 \times (1 + 0.07 \times 3) = \$400 \times (1 + 0.21) = \$400 \times 1.21 = \$484.$$

The next example shows how a calculator with algebraic logic may be used to do calculations.

EXAMPLE 6.3 What is the simple interest on a \$500 loan at 12% from June 6 until October 12 (in a nonleap year)?

History

The length of the "year" has varied considerably over recorded history. The earliest Babylonian year was determined by the occurrence of the lunar eclipse, which happens approximately every six months. Around 1800 B.C. the Assyrians had a year of exactly 360 days divided into 12 equal months, just as is used in computing ordinary interest.

Solution

To find the time period as measured in years, we first add up the days: $(30 - 6) + 31 + 31 + 30 + 12 = 128$. Since the problem concerns a nonleap year, there are 365 days in the year, so the time period is $\frac{128}{365}$ of a year. Next compute the interest: 500 [×] 0.12 [×] 128 [÷] 365 [=] [21.041096], or \$21.04.

The preceding interest calculation may seem complicated. Since all the arithmetic can be done with a calculator, the hardest part is determining the number of days in the time period. To simplify the calculation, a type of simple interest called ordinary interest was created. **Ordinary interest** is based on two accepted conventions: (i) each month is assigned 30 days and (ii) a year is assigned 360 days. If we had used ordinary interest, the time period in the last example would have been 4 months from June 6 to October 6 plus 6 days from October 6 to October 12. That gives us $(4 \times 30) + 6 = 126$ days out of a 360-day year. The interest is then

500 [×] 0.12 [×] 126 [÷] 360 [=] [21], or \$21.

Since ordinary interest makes individual days more expensive, it is not appropriate for computing simple interest on a short time period in days. Because of the widespread use of pocket calculators, the simplification of dividing by 360 instead of 365 or 366 is unimportant. But when the time period is in months, as is often the case for short-term borrowing, *it is standard to treat all months as exactly $\frac{1}{12}$ of a year.*

EXAMPLE 6.4 What is the ordinary interest on a \$500 loan at 12% from June 1999 through September 2000? What is the future value of the loan?

(*Note:* Our use of *through* will mean that both June and September are included.)

Solution

Counting all the months involved as complete months, there are 16 months from June 1999 through September 2000. Hence we have

$$\$500 \times 0.12 \times \frac{16}{12} = \$80.$$

The future value of the loan is the sum of the principal and the interest or

$$F = \$500 + \$80 = \$580.$$

Tidbit

This problem was posed almost 800 years ago by Leonardo of Pisa, who used the pen name Fibonacci. The sequence is known as the Fibonacci sequence.

Compound Interest

A problem dating back to the 1200s asks how many rabbits you would have if you started with just two and let them reproduce. Assume that a pair of rabbits produces a pair of offspring each month and that each pair of rabbits produces their first offspring at age 2 months. Month by month the number of pairs of rabbits is

Month	1	2	3	4	5	6	7	8	9	10	11	12	13	14	15	16
Pairs of Rabbits	1	2	3	5	8	13	21	34	55	89	144	233	377	610	987	1597

After just 16 months, you have almost 1600 pairs of rabbits!

Like the young rabbits becoming part of the adult breeding stock, reinvesting interest income makes the amount of your money grow faster. This phenomenon, called **compound interest**, is a powerful way to make money grow. To calculate compound interest you need the same information that is used to calculate simple interest, namely, the principal, the interest rate, and the time period. In addition, you need the **compounding period**. The principal is the initial amount deposited in the account. The amount in the account at any time will be called the **balance**. (So the principal is the **initial balance**.) For each compounding period, calculate simple interest on the balance in the account at the start of the compounding period. At the end of the compounding period, add the interest to the balance. This becomes the principal for the next compounding period. Typical compounding periods are one month, three months, and six months; in these cases, ordinary interest computations are used instead of the more laborious simple interest method.

EXAMPLE 6.5 Given \$1000 principal, a 10% interest rate,and a six-month compounding period, find the balance after two years.

Solution

Ordinary interest at 10% $(=0.10)$ for six months $(\frac{1}{2}$ year$)$ on \$1000 is

$$\$1000 \times 0.10 \times \tfrac{1}{2} = \$50.$$

At the end of the six months this \$50 is added to the balance, so the new balance is \$1050. Ordinary interest for the next six months at 10% on \$1050 is

$$\$1050 \times 0.10 \times \tfrac{1}{2} = \$52.50.$$

The new balance at the end of one year is $\$1050 + \$52.50 = \$1102.50$. Ordinary interest for six months at 10% on \$1102.50 is

$$\$1102.50 \times 0.10 \times \tfrac{1}{2} = \$55.13.$$

Thus the new balance at the end of $1\frac{1}{2}$ years is \$1157.63. Ordinary interest for six months at 10% on \$1157.63 is

$$\$1157.63 \times 0.10 \times \tfrac{1}{2} = \$57.88.$$

The new balance at the end of two years is $\$1157.63 + \$57.88 = \$1215.51$. ■

Notice that in the preceding example, simple interest at 10% on \$1000 for two years is

$$\$1000 \times 0.10 \times 2 = \$200,$$

so compound interest gave \$15.51 more than simple interest. Compound interest is always larger than simple interest at the same rate, and the longer the period, the more striking the difference.

Computing Compound Interest by a Formula

Calculating compound interest is not as complicated as it seems. The important thing to realize is that the balance at the end of a compounding period can be computed by simply multiplying the balance at the start of the compounding period by

$$1 + (\text{interest rate} \times \text{compounding period}),$$

where the interest rate is expressed as a decimal, and the compounding period is expressed in years. For the previous example, the interest rate is 10% and the compounding period is six months. Thus you multiply by

$$1 + \left(0.10 \times \tfrac{1}{2}\right) = 1.05.$$

Since the two years in the example represents four compounding periods, we multiply the initial balance by 1.05 four times to get the final balance:

1000 [×] 1.05 [×] 1.05 [×] 1.05 [×] 1.05 [=] [1215.50625]

which rounds to \$1215.51.

Repeated multiplication can be written using power or exponential notation. So

$$1.05 \times 1.05 \times 1.05 \times 1.05 = 1.05^4,$$

and the final balance in the example can be obtained by using the [x^y] (or [y^x]) key of a scientific calculator as follows:

1.05 [x^y] 4 [×] 1000 [=] [1215.50625]

and then rounding off to the nearest penny.

> ***COMPOUND INTEREST FORMULA***
>
> **If P represents the principal, r the interest rate expressed as a decimal, m the number of equal compounding periods (in a year), and t the time in years, then the final balance, F, is given as**
>
> $$F = P \times \left(1 + \tfrac{r}{m}\right)^{mt}.$$

The final balance, F, in the preceding compound interest formula represents the **future value** of the principal after t years of compound interest at the interest rate r, assuming m compounding periods per year.

EXAMPLE 6.6 Find the final balance in the following savings accounts having an initial balance of \$2457.

(a) simple interest at 4.5% for three years
(b) interest at 4.5% compounded every four months for three years
(c) interest at 4.5% compounded monthly for three years

Solution

(a) 2457 [×] (1 [+] 0.045 [×] 3) [=] [2788.695]
(*Note:* The 0.045 is multiplied by 3, not the 1 + 0.045.)

(b) When compounding every four months, $m = 3$.

2457 [×] (1 [+] 0.045 [÷] 3) [x^y] (3 [×] 3) [=] [2809.30917]

(*Note:* By the compound interest formula, the 0.045 was divided by 3 since there are 3 four-month periods per year.)

(c) When compounding monthly, $m = 12$.

2457 [×] (1 [+] 0.045 [÷] 12) [x^y] (12 × 3) [=] [2811.416924]

There are two important observations that can be made with respect to the solution of Example 6.6.

1. Most scientific calculators use algebraic logic. If yours does not, you will have to make adjustments in your keystrokes. For example, in (b), you might try the following:

   0.045 [÷] 3 [+] 1 [=] [x^y] (3 [×] 3) [=] [×] 2457 [=] [2809.30917]

2. The correct use of parentheses is crucial.

Effective Annual Rate—The Effect of the Compounding Period

To compare different savings plans, you need to have a common basis for making the comparisons. The **effective annual rate (EAR)**, or **annual yield**, provides such a basis. The effective annual rate is the simple interest rate that would have earned the same amount of interest in one year. To find the effective annual rate, the easiest approach is to compute what happens to \$100 over 1 year.

Tidbit

The characteristic of being convertible to cash is called liquidity. Often, to obtain a high yield you must sacrifice liquidity.

EXAMPLE 6.7 What is the effective annual rate on a savings account paying 12% with a compounding period of three months?

Solution

Starting with \$100 principal, we use the compound interest formula to find the balance: $\$100 \times (1 + \frac{0.12}{4})^4 = \$112.55$. Since the balance in the account has increased by \$12.55 over the year, the effective annual rate is 12.55%. This compares to the stated rate of 12%.

History

In 1777 Jacob DeHaven loaned General George Washington \$50,000 in gold and \$400,000 worth of supplies. When offered Continental money as repayment, DeHaven refused, insisting on gold. He was never repaid. In 1989 members of DeHaven's family sued for repayment. Because of compound interest the amount involved has increased to about \$150 billion.

In this example, the dollar increase in the account translates exactly to the effective annual rate in percent; this is the reason for considering a \$100 initial balance. Financial institutions are required to inform consumers of the effective annual rate of their various savings options to help consumers make informed decisions. The effective annual rate should be an important factor in your choice of savings account. Some accounts do not allow you to withdraw your money at will, a provision that should also be an important consideration.

In Example 6.7, \$100 was used for convenience in our computation. More formally, the effective annual rate for each dollar on deposit is given by $EAR = F - 1$, where F is the final balance, with interest, generated by \$1.

Substituting $P = 1$ (for \$1), $t = 1$ (for 1 year), and $EAR = F - 1$ in the compound interest formula, we obtain the following formula. The resulting decimal can be directly translated as a percentage.

> ***EFFECTIVE ANNUAL RATE FORMULA***
>
> If EAR represents the effective annual rate, r the interest rate expressed as a decimal, and m the number of equal compounding periods (in a year), then
>
> $$EAR = (1 + \tfrac{r}{m})^m - 1.$$

There are two factors that determine the effective annual rate: the interest rate, r, and the number of compounding periods, m. The effective annual rate is always at least as large as the interest rate. How much larger it is depends on how often the interest is compounded. Table 6.1 shows the effective annual rate of various combinations of interest rates and compounding periods.

TABLE 6.1
EFFECTIVE ANNUAL RATE TABLE

	Compounding Periods						
Percent	*1 (Simple)*	*2 (Semi-ann)*	*4 (Quarterly)*	*12 (Monthly)*	*365 (Daily)*	*1000*	*Continuously Compounded*
5	5.00000	5.06250	5.09453	5.11619	5.12675	5.12698	5.12711
6	6.00000	6.09000	6.13636	6.16778	6.18313	6.18346	6.18365
7	7.00000	7.12250	7.18590	7.22901	7.25010	7.25056	7.25082
8	8.00000	8.16000	8.24322	8.29995	8.32776	8.32836	8.32871
9	9.00000	9.20250	9.30833	9.38069	9.41621	9.41699	9.41743
10	10.00000	10.25000	10.38129	10.47131	10.51558	10.51654	10.51709
11	11.00000	11.30250	11.46213	11.57188	11.62596	11.62713	11.62781
12	12.00000	12.36000	12.55088	12.68250	12.74746	12.74887	12.74969
13	13.00000	13.42250	13.64759	13.80325	13.88020	13.88188	13.88284
14	14.00000	14.49000	14.75230	14.93420	15.02429	15.02625	15.02738
15	15.00000	15.56250	15.86504	16.07545	16.17984	16.18212	16.18342

In the table, the number of compounding periods per year is indicated in the top row, and the interest rates are in the left column. As you read across each row, you can see that the effective annual rate keeps getting larger and larger as the number of compounding periods gets larger. However, there is a limit to this growth, and the yield levels out, approaching the well-defined limit in the far right column. For the limiting value, the number of compounding periods in a year is infinite, and the interest is said to be **continuously compounded**. If r is the interest rate expressed as a decimal, the effective annual rate for continuously compounded interest is

$$e^r - 1,$$

which is easily computed on any calculator with the $\boxed{e^x}$ key. (*Note:* The value of e is approximately 2.72.)

EXAMPLE 6.8 Use (a) the table and (b) a calculator to find the effective annual rate of an account paying 5% interest compounded monthly.

Solution

(a) In the intersection of the row for 5% interest and the column for monthly compounding, we find 5.11619, so the effective annual rate is 5.12%.

(b) $EAR = (1 \boxed{+} 0.05 \boxed{\div} 12) \boxed{x^y} 12 \boxed{-} 1 \boxed{=} \boxed{0.0511619}$, or, rounding up, 5.12%.

INITIAL PROBLEM SOLUTION

You discover that you are the only direct descendant of a man who loaned the Continental Congress $1000 in 1777. However, he was never repaid nor have any of his descendants received repayment. You think it is about time to get the family money back. How much should you demand from the U.S. government? Use an interest rate of 6% and a compounding period of three months.

Solution

Historically, a typical interest rate has been approximately 6% and a compounding period of three months has also been typical. To apply the compound interest formula we replace P by \$1000, r by 0.06, and m by 4. The time period is from 1777 to 2000, which is $2000 - 1777 = 223$ years, so we replace t in the formula by 223 and get

$$\$1000 \times [1 + 0.06 \div 4]^{(223\times 4)} = \$1000 \times 1.015^{892} = \$585{,}746{,}479.$$

You should demand $585,746,479 (but expect to get much less).

Problem Set 6.1

Problems 1 through 12

P is the principal and r is the annual rate as a percent.

1. Find simple interest.
   - **(a)** $P = \$600$ $r = 7\%$ $t = 3$ years
   - **(b)** $P = \$400$ $r = 12\%$ $t = 5$ years
   - **(c)** $P = \$1235$ $r = 7\%$ $t = 10$ years
2. Find the simple interest.
   - **(a)** $P = \$525$ $r = 5\%$ $t = 2$ years
   - **(b)** $P = \$300$ $r = 3\%$ $t = 5$ years
   - **(c)** $P = \$7934$ $r = 4.15\%$ $t = 8$ years
3. Find ordinary interest.
   - **(a)** $P = \$800$ $r = 6\%$ $t = 36$ months
   - **(b)** $P = \$1400$ $r = 12\%$ $t = 30$ months
   - **(c)** $P = \$1235$ $r = 7.5\%$ $t = 20$ months
4. Find ordinary interest.
   - **(a)** $P = \$525$ $r = 5\%$ $t = 48$ months
   - **(b)** $P = \$300$ $r = 3\%$ $t = 40$ months
   - **(c)** $P = \$7934$ $r = 4.15\%$ $t = 33$ months
5. Find the simple interest on a $650 loan at 6% from January 10 to November 16 (in a nonleap year).
6. Find the simple interest on a $1200 loan at 10% from February 10 to June 28 (in a leap year).
7. Find the simple interest on a $1600 loan at 7.5% from February 12 to November 6 (in a leap year).
8. Find the simple interest on a $2000 loan at 9% from March 15 to October 8 (in a nonleap year).
9. Find ordinary interest.
   - **(a)** $P = \$800$, $r = 6\%$, from 6/97 through 2/99
   - **(b)** $P = \$950$, $r = 7.5\%$, from 10/96 through 3/99
10. Find ordinary interest.
    - **(a)** $P = \$700$, $r = 5\%$, from 1/96 through 3/99
    - **(b)** $P = \$385$, $r = 8.5\%$, from 3/96 through 2/99
11. Find ordinary interest.
    - **(a)** $P = \$2350$, $r = 5.6\%$, from 7/96 through 9/99
    - **(b)** $P = \$7200$, $r = 5\%$, from 1/96 through 3/99
12. Find ordinary interest.
    - **(a)** $P = \$7500$, $r = 6.1\%$, from 5/92 through 11/98
    - **(b)** $P = \$1250$, $r = 5\%$, from 12/92 through 3/99
13. Find the final balance for a $2000 loan at 5% simple interest for three years.
14. Find the final balance for a $3579 loan at 7.25% simple interest for five years.
15. Find the final balance for a $3000 loan at 6.5% simple interest for five years.
16. Find the final balance for a $4550 loan at 7.5% simple interest for three years.
17. Find the final balance for a $2575 loan at 5.75% ordinary interest for 40 months.
18. Find the final balance for a $2400 loan at 6.25% ordinary interest for 28 months.
19. Given $1500 principal, a 12% interest rate, and a three-month compounding period, find the balance after one year. Calculate the new balance at the end of each compounding period.
20. Given $2000 principal, an 8% interest rate, and a six-month compounding period, find the balance after one year. Calculate the new balance at the end of each compounding period.

21. Which is the best deal over three years?
   (a) investing at 5% compounded annually
   (b) investing at 4.95% compounded semiannually
   (c) investing at 4.9% compounded monthly.

22. Which is the best deal over five years?
   (a) investing at 8% compounded annually
   (b) investing at 7% compounded monthly
   (c) investing at 6.8% compounded continuously

23. What is the effective annual rate to the nearest hundredth of a percent on an account paying 8% compounded every two months?

24. What is the effective annual rate on an account paying 7% compounded every five days?

25. John's parents agree to loan him $5000 on the condition that he pays them $6000 at the end of five years. What simple rate of interest is John paying to his parents?

26. A loan of $4000 is made with the condition that if it is paid back within a year, a simple interest rate of 8% will be charged.
   (a) If the loan was made on June 10, 1998, and repaid on February 17, 1999, what amount of interest should be charged?
   (b) What would be the amount charged if ordinary interest is used?

27. An investor bought Signal Microchips stock for $16 a share. If the annual dividend is $1.50 per share and the stock was valued at $21.50 per share at the end of one year, what simple interest rate (including the dividend) was earned on the investment?

28. Marcia loaned $3000 to a friend for 90 days at 12% simple interest. After one month, she sold the note to a third party for the original amount of $3000. What interest rate did the third party receive? Use ordinary interest in the calculation.

29. A student has a savings account earning 6% simple interest. She must pay $1500 for the first semester tuition by September 1 and $1500 for the second semester by January 1. How much must she have on hand at the end of the summer (by September 1) in order to pay the first semester tuition on time and still have the remainder of her funds grow to $1500 by January 1? Use ordinary interest in the calculation.

30. An investor owns several apartment buildings. The taxes on these buildings total $30,000 for the year and are due before April 1. The late fee is 1/2% per month up to six months, at which time more severe penalties will be assessed. If the investor has $30,000 available on March 31, will he save money by paying the taxes on time or by investing the money at 8% simple interest and paying the taxes and late fee on September 30?

31. What is the effective annual rate equivalent to a rate of 8%, compounded continuously?

32. If money is invested at 6.5%, compounded continuously, what is the effective annual rate?

Problems 33 through 36

Make a guess and check your answer. Then adjust your guess and check again until you have the right answer.

33. How long (to the nearest year) would $1200 have to be invested at 8%, compounded annually, to amount to a total of $1925?

34. How long will it take for $2000 to double at 10% interest compounded annually?

35. How long will it take for $500 to grow to $2500 at 15% compounded semiannually?

36. What amount should be deposited to yield $2000 if it is compounded at 8% annually for five years?

37. A child's parents want to have $40,000 in 15 years to pay for her college education. What amount should they deposit if they can earn 8% compounded annually?

38. An IRA (Individual Retirement Account) allows a saver to save tax-deferred (i.e., taxes are paid when the money is withdrawn). If a person deposits $5000 in an IRA earning 5% compounded quarterly, how much will the account be worth in 20 years?

39. How much money would you need to have saved to earn $25,000 interest per year (Use simple interest in the calculation)
   (a) if you could get 4% interest?
   (b) if you could get 6% interest?
   (c) if you could get 10% interest?

40. How much money would you need to have saved to earn $100,000 interest per year (Use simple interest in the calculation)
   (a) if you could get 5% interest?
   (b) if you could get 8% interest?
   (c) if you could get 12% interest?

Problems 41 and 42

If an investment grows in value from P_0 to P_1 over a period of N years, the equivalent annual rate of growth is

$$\left(\frac{P_1}{P_0}\right)^{1/N} - 1,$$

which when converted to a percentage gives the effective annual rate.

41. Suppose the value of your mutual fund account increased from $3000 to $4500 over five years. Assuming all dividends were reinvested, what was the effective annual rate?

42. Suppose you paid $10,000 for bonds that pay 12% interest. After three years you sell the bonds. Since it is prior to maturity and interest rates have changed, you only receive $9500. What is the effective annual rate for this investment?

Extended Problems

43. Investigate the interest rates available to you from the following sources.
  (a) savings accounts in banks, credit unions, and savings and loans
  (b) certificates of deposit in banks, credit unions, and savings and loans
  (c) United States savings bonds
  (d) money market funds
  (e) Treasury bills and notes

44. Research the relationship between short-term and long-term interest rates. In particular, is one usually higher than the other?

45. Research the historical record on interest rates. Discuss the important factors that seem to influence interest rates.

46. What are "junk bonds"? Why would anyone buy a junk bond?

The compounding of interest is an example of what is known as "exponential growth." This concept has important applications when looking at the growth of human, or other, populations; in this case, our population consists of money. One important way to assess the effects of growth is called the doubling time. Specifically, the doubling time is the length of time needed for an amount to double in value. When an amount is continuously compounded, the compound interest formula may be written as $F = Pe^{rt}$, where $e \approx 2.7183$, r is the rate expressed as a decimal, and t is the time in years. If the amount, F, is to be twice that of P, then $e^{rt} = 2$. Since $e^{0.6932} \approx 2$ (check this on your calculator), we see that the amount doubles when $rt = 0.6932$. Equivalently, the amount doubles when $t = 0.6932/r$. For example, if the rate is 6%, the doubling time is $\frac{0.6932}{0.06} \approx 69/6$ or 11.5 years. To find the doubling time exactly requires the use of logarithms, but the method given in the approximation is satisfactory for most purposes. Since we saw that compounding monthly gives values close to those of compounding continuously, the method also gives good approximations for most periodic compounding situations. To make things even easier, 72 is often used instead of 69 because it has many more integer factors than 69. This method of approximation has a name: the rule of 72. Using the rule of 72, we would say that an amount that is being compounded at 6% will double within 12 years (72/6), while an amount being compounded at 8% will double within 9 years.

47. Find the doubling time for a deposit of $2500 earning 9% compounded monthly. Check your result by using the compound interest formula.

48. Find the doubling time if an investment earns 7.2% compounded monthly. Check your result by using the compound interest formula.

49. Find the doubling time if an investment earns 4.5% compounded monthly. Check your result by using the compound interest formula.

50. How many years will it take for an investment to quadruple in value if it is being compounded monthly at 6%?

6.2 LOANS

INITIAL PROBLEM

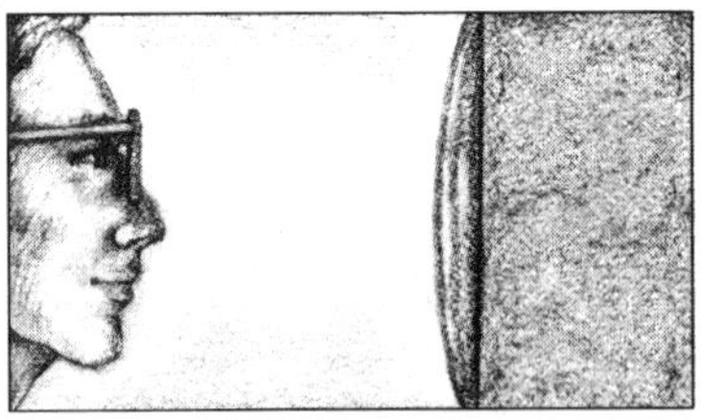

Suppose you can rent a $500 television for $30 a month and after 24 months you own it. Is this a good idea, or would it be better to charge it on your credit card and pay off that credit card account at a rate of $30 a month?

Having discussed the ways in which interest is charged and paid, we will now consider, in more detail, the most common forms of loans. In this section, we look at loans that are based on simple interest.

Simple Interest Loans

The interest on a **simple interest loan** is simple interest on the amount currently owned. Credit card accounts are a common example. Each month the bank or charge card company charges simple interest, called the **finance charge**, based on the balance owed. Many credit cards also have a grace period during which no

Tidbit

Conditions on most credit cards have become fairly complex. Many national credit cards have a variety of restrictions and conditions reflecting differences in state laws.

interest is charged if full payment is received by the payment due date. The most common method for calculating finance charges uses the average daily balance. When the **average daily balance** is used, a cardholder is only charged for the actual number of days each amount owed was carried on the bill. This method converts the annual percentage rate to a daily interest rate.

To calculate the average daily balance, you determine the outstanding balance for each day, and divide the sum of these daily balances by the number of days in the monthly billing period. Any payments or other credits are subtracted from the previous day's balance as they occur. In general, the monthly statement includes the current month's charges, any unpaid balance, and finance charges.

EXAMPLE 6.9 The statement for Bob Chargeit's credit card account shows the following activity:

June 12	auto repair	\$45.60
June 18	payment	\$150.00
June 22	gasoline	\$20.00
July 3	paint	\$78.50

Find the average daily balance, the finance charge, and new balance if the billing period is from June 10 through July 9, inclusive, the previous balance was \$287.84, and the annual percentage rate is 21%.

Solution

To find the average daily balance, we use the balance for each day in the billing period and the number of days the balance was in effect, as shown in the following table. Each balance is multiplied by the number of days it was in effect; we multiply 6×333.44 rather than add 333.44 for six different days. The values are added, and the total is divided by the number of days in the billing period.

Time Period	Days	Daily Balance
June 10–June 11	2	\$287.84
June 12–17	6	\$287.84 + \$45.60 = \$333.44
June 18–June 21	4	\$333.44 − \$150.00 = \$183.44
June 22–July 2	11	\$183.44 + \$20.00 = \$203.44
July 3–July 9	7	\$203.44 + \$78.50 = \$281.94

$$\text{average daily balance} = \frac{2(287.84) + 6(333.44) + 4(183.44) + 11(203.44) + 7(281.94)}{2 + 6 + 4 + 11 + 7}$$

$$= \frac{7521.50}{30} = \$250.72$$

The finance charge is the simple interest on the average daily balance using a daily interest rate. For an annual percentage rate of 21%, the daily percentage rate is 0.057534%, which is $\frac{21\%}{365}$. We then use the simple interest formula $I = Prt$, with the rate as a decimal and time given in terms of days:

$$\text{finance charge} = 250.72 \times \frac{0.21}{365} \times 30 = 4.327496 = \$4.33.$$

The new balance on the account will be the sum of the ending balance and the finance charge:

$$\text{new balance} = \$281.94 + \$4.33 = \$286.27.$$

The particulars of how the balance on a credit card is determined vary. But what happens if you allow a balance to recur month after month? The answer is that you end up paying compound interest to the credit card company!

Add-On Interest

Sometimes businesses offer to finance the purchase of furniture, appliances, or automobiles with monthly payments using what is called **add-on interest**. To find the monthly payment for such a purchase, calculate simple interest at the annual interest rate over the length of the loan agreement. Then divide the sum of the purchase price and the interest into equal monthly payments. In other words, you will pay the lender the same future value of a simple interest loan as calculated in Section 6.1, but instead of paying the entire amount in a lump sum at the end of the loan, you will pay it in equal monthly installments over the life of the loan.

EXAMPLE 6.10 A \$1200 motorcycle is financed over a two-year period with 12% add-on interest. Find the monthly payment.

Solution

Simple interest at 12% over two years on \$1200 is

$$\$1200 \times 0.12 \times 2 = \$288.$$

Adding the \$288 interest and the \$1200 principal yields \$1488, which must be paid in 24 equal payments. So the monthly payment is \$1488 ÷ 24 = \$62.00.

Annual Percentage Rate

In 1969, Congress passed the Consumer Credit Protection Act, which is usually known as the Truth-in-Lending Act, requiring lenders to compute and disclose the annual percentage rate of any loan that is not a simple interest loan. The **annual percentage rate (APR)** is the interest rate on an amortized loan that would require the same payments (see section 6.3). Notice that add-on interest loans charge interest on the *entire* principal over the life of the loan even though you don't have use of all of the money during that period. Computing the annual percentage rate for add-on interest loans is difficult to do from formulas. Instead, tables showing the APRs for different combinations of rates and loan periods are used.

EXAMPLE 6.11 Find the annual percentage rate for the purchase of the motorcycle in the previous example, namely, a 12% loan over two years.

Solution

Table 6.2 gives some sample APRs.

TABLE 6.2
APR TABLE OF ADD-ON INTEREST LOANS

	Length of the Loan in Years		
Nominal Interest Rate	*1*	*2*	*5*
6	10.9	11.1	10.8
8	14.5	14.7	14.1
10	18.0	18.2	17.3
12	21.5	21.6	20.3

Since it was an add-on interest loan with a nominal rate of 12% and length of the loan was two years, the APR is 21.6%, the intersection of the "12" row and the "2 year" column in the APR table.

The APR table can be used to estimate the annual percentage rate for loans. Observe that in the 10% row, the add-on interest rates are 18.0, 18.2, and 17.3. Comparing these rates to 10 shows that the annual percentage rate for an add-on interest loan is approximately *1.8 times* the nominal interest rate! This is a good estimate for most interest rates and time periods. Since there is no simple formula for the APR, you should use the estimate of multiplying by 1.8 when an APR table is unavailable.

EXAMPLE 6.12 Find the approximate annual percentage rate for an add-on interest loan of 20% for one year.

Solution

We use the factor of 1.8 for approximations: $20\% \times 1.8 = 36\%$.

Rent-to-Own

Since the passage of the Truth-in-Lending Act forced some lenders to reveal the extremely high annual percentage rates they charge, add-on interest loans have become rare. The replacement for add-on interest is the rent-to-own transaction, in which you rent an item you cannot afford to buy outright. After a contracted number of payments, the item becomes yours. Of course, the rental may be for a shorter period of time, and the item is then returned. The effect of a rent-to-own transaction is the same as buying on credit, but technically it is not a credit purchase and is thus not covered by the Truth-in-Lending Act.

For comparison shopping on rates, you can still treat a rent-to-own transaction as a loan and compute the annual percentage rate. Do this as follows: Find the total of all the payments required to buy the item and subtract the best purchase price available at an ordinary retail store; the difference is essentially the add-on interest. Find the simple interest rate that would have to be charged on the retail purchase price to amount to the add-on interest; then multiply by 1.8 to get the approximate annual percentage rate.

EXAMPLE 6.13 Suppose you can rent-to-own a \$500 television for 24 monthly payments of \$30. Estimate the annual percentage rate you would be paying on rent-to-own.

Solution

The payments total \$30 $\times$ 24 = \$720. You will be paying \$720 − 500 = \$220 as add-on interest over two years, or \$110 per year. The simple interest rate needed for that amount of interest is 110/500 = 0.22 = 22%. Since we do not have a table that includes the APR for a loan with nominal interest rate of 22% and term of two years, we must make an estimate by multiplying the nominal rate by 1.8. So we estimate that the annual percentage rate is approximately 1.8 $\times$ 22% $\approx$ 40%!

The annual percentage rate on rent-to-own transactions will usually turn out to be very high, partly due to the fact that these transactions are, in general, a higher risk for the merchant. If possible, you would be well advised to go elsewhere to borrow the money needed to make the ordinary retail purchase.

INITIAL PROBLEM SOLUTION

Suppose you can rent a \$500 television for \$30 a month and after 24 months you own it. Is this a good idea, or would it be better to charge it on your credit card and pay off that credit card account at a rate of \$30 a month?

Solution

The annual interest rate of the rent-to-own option is about 40%, as shown in Example 6.13. Credit card accounts generally have annual percentage rates around 20% or lower. Since even the nominal rate of 22% of the rent-to-own option is higher than 20%, the credit card is clearly the way to go.

Problem Set 6.2

Problems 1 through 4
Find the finance charge on a credit card account for the given conditions. When billing periods are given, the use of *through* means the beginning and ending dates are both included.

1. (a) Average Daily Balance = \$255.00,
Annual Percentage Rate = 12.9%,
Billing Period = 30 days

(b) Average Daily Balance = \$425.80,
Annual Percentage Rate = 14.9%,
Billing Period = 31 days

2. (a) Average Daily Balance = \$183.65,
Annual Percentage Rate = 16.9%,
Billing Period = 31 days

(b) Average Daily Balance = \$194.85,
Annual Percentage Rate = 16.5%,
Billing Period = 30 days

3. (a) Average Daily Balance = \$315.42,
Annual Percentage Rate = 14.9%,
Billing Period: May 15 through June 14

(b) Average Daily Balance = \$275.65,
Annual Percentage Rate = 15.9%,
Billing Period: March 11 through April 10

4. (a) Average Daily Balance = \$583.27,
Annual Percentage Rate = 17.9%,
Billing Period: April 11 through May 10

(b) Average Daily Balance = \$224.85,
Annual Percentage Rate = 16.5%,
Billing Period: July 15 through August 14

Problems 5 through 10
Find the finance charge and new balance for each credit card account.

5. Ending Balance = \$320.50,
Average Daily Balance = \$275.00,
Annual Percentage Rate = 18.9,%
Billing Period = 30 days

6. Ending Balance = \$485.88,
Average Daily Balance = \$325.80,

Annual Percentage Rate = 14.9%,
Billing Period = 31 days

7. Ending Balance = \$147.85,
Average Daily Balance = \$155.00,
Annual Percentage Rate = \$15.9%,
Billing Period = 30 days

8. Ending Balance = \$227.54,
Average Daily Balance = \$215.80,
Annual Percentage Rate = 18.9%,
Billing Period = 31 days

9. Ending Balance = \$135.92,
Average Daily Balance = \$105.00,
Annual Percentage Rate = 16.9%,
Billing Period = 31 days

10. Ending Balance = \$362.00,
Average Daily Balance = \$325.80,
Annual Percentage Rate = 14.9%,
Billing Period = 30 days

Problems 11 through 14

The activity in a credit card account is given for one month.

11. Billing period: 10/11 through 11/10; previous balance: \$165.45; annual interest rate: 12.9%.

October 18	Payment	\$100.00
October 25	Restaurant	\$ 28.90
November 5	Software	\$ 85.64

(a) Find the average daily balance and the finance charge.
(b) What is the new account balance on November 11?

12. Billing period: 9/11 through 10/10; previous balance: \$385.56; annual interest rate: 14.9%.

September 15	Payment	\$200.00
September 22	Bookstore	\$ 42.85
October 2	Clothes	\$192.93

(a) Find the average daily balance and the finance charge.
(b) What is the new account balance on October 11?

13. Billing period: 6/11 through 7/10; previous balance: \$225.85; annual interest rate: 14.9%.

June 20	Shoes	\$ 79.95
June 25	Payment	\$125.00
June 28	Books	\$ 34.65
July 5	Radio	\$ 69.50

(a) Find the average daily balance and the finance charge.
(b) What is the new account balance on July 11?

14. Billing period: 3/11 through 4/10; previous balance: \$95.15; annual interest rate: 15.8%.

March 15	Clothes	\$113.50
March 20	Hardware	\$ 52.93
March 20	CDs (music)	\$ 28.67
March 22	Payment	\$175.00

(a) Find the average daily balance and the finance charge.
(b) What is the new account balance on April 11?

Problems 15 through 20

Find the monthly payment on contracts involving add-on interest. Remember that the total of the monthly payments for an add-on interest loan is computed using the simple interest loan formula, $F = P(1 + rt)$.

15. A \$675 stereo is financed over a two-year period with 15% add-on interest. What is the monthly payment?

16. Louise buys a new bike for \$375 and pays for it over two years with 13.5% add-on interest. What is her monthly payment?

17. Sean's new snowboard cost him \$425. He financed it over an 18-month period with 15% add-on interest. What is the monthly payment?

18. A used car that sold for \$5775 is financed for 30 months with 9.5% add-on interest. What is the monthly payment?

19. Ron and Shannon buy a new refrigerator for \$755. The purchase is financed over 30 months with 10.5% add-on interest. What is their monthly payment?

20. What monthly payment is required for a \$425 set of skis that are financed for one year with 13.5% add-on interest?

21. What is the original principal for a simple interest loan that had a total value of \$4340 after three years at 8%?

22. Jerry's grandfather loaned him money to help buy a car. Two years later, when Jerry paid back the loan, his grandfather only charged him 3.5% simple interest on the loan. If Jerry paid his grandfather \$2568, how much had he borrowed?

23. What was the original purchase price (to the nearest dollar) of a television purchased with 20 monthly payments of \$48.03, which include add-on interest of 12.5%?

24. Lucinda purchased a new watch from a jewelry store and paid for her purchase with 12 monthly payments of \$34.53 based on add-on interest of 10.5%. What was the purchase price?

25. Find the APR for each of the following nominal add-on interest rates.

(a) 12% for five years **(b)** 6% for two years
(c) 10% for one year **(d)** 8% for two years

26. Find the APR for each of the following nominal add-on interest rates.

(a) 12% for one year **(b)** 6% for five years
(c) 10% for two years **(d)** 8% for five years

27. If a \$600 television is purchased on a rent-to-own agreement at \$32 a month for 24 months, what amount is paid above and beyond the stated purchase price?

28. A dining room set at a rent-to-own store can be rented for $65 a month (a deposit and a minimum of three months are required). If the set has a suggested retail value of $1250 and you own the set after 24 months, how much extra do you pay under this arrangement?

Problems 29 through 32

Find the annual add-on interest rate being used.

29. **(a)** Principal = $500, term = 1 year, monthly payment = $46.47
    (b) Principal = $750, term = 2 years, monthly payment = $37.50
30. **(a)** Principal = $925, term = 2 years, monthly payment = $48.18.
    (b) Principal = $1500, term = 3 years, monthly payment = $58.17
31. **(a)** Principal = $600, term = 18 months, monthly payment = $40.58
    (b) Principal = $450, term = 20 months, monthly payment = $28.69
32. **(a)** Principal = $1150, term = 30 months, monthly payments = $50.31
    (b) Principal = $1800, term = 36 months, monthly payment = $75.35
33. Estimate the APR for a loan with add-on interest of 12.5%
34. Estimate the APR for a loan with add-on interest of 10.5%
35. If a $600 television is purchased on a rent-to-own agreement at $32 a month for 24 months, what nominal rate of interest is being charged if you consider the extra charges as add-on interest? What is the estimated APR?
36. If a dining room set with a suggested retail value of $1250 is purchased under a rent-to-own agreement of $65 a month for 24 months, what nominal rate of interest is being charged if you consider the extra charges as add-on interest? What is the estimated APR?
37. If a $1500 home entertainment center is purchased on a rent-to-own agreement at $70 a month for 30 months, what nominal rate of interest is being charged if you consider the extra charges as add-on interest? What is the estimated APR?
38. A living room set with a suggested retail price of $1050 is available at two rent-to-own stores. One store has the set at a rent-to-own rate of $55.56 a month for 24 months; at the other store, the rent-to-own rate is $63 a month for 20 months. What are the annual interest rates being charged when you consider the extra charge as add-on interest? What are the estimated APRs?
39. Ted buys a used car priced at $2500 with no down payment and makes payments of $125 a month for two years. What is the annual interest rate if the finance charges are figured as add-on interest? What is the estimated APR?
40. A student buys a new stereo system that costs $975 by agreeing to make monthly payments for three years to cover the cost of the set and 12% add-on interest.
    (a) What is the amount of the monthly payment?
    (b) What is the estimated APR?
41. Jamie decides to get a new table and chairs and finds a suitable set at a rent-to-own dealer. The terms are $32 a month, with ownership after 30 months. Then Jamie finds the same basic set for sale at $629 at a local furniture store. What is the approximate annual percentage rate that Jamie could pay for a conventional loan and still have payments that were no more than $32 a month?
42. Jorge is either going to buy a new stereo system or get the same system through the rent-to-own dealer. He can buy the set for $796; at the rent-to-own dealer he would pay $42.50 for 24 months. What is the approximate annual percentage rate that Jorge could pay for a conventional loan and still have payments that were no more than $42.50 a month?

Extended Problems

43. Investigate the interest rates available on bank credit cards. Is there any reason a person would choose a card with a high interest rate?
44. Compare the terms at a rent-to-own store with the available retail price for an appliance of interest to you. What is the approximate annual percentage rate if the rent-to-own arrangement is treated like a loan with add-on interest?
45. Research the history of consumer credit. Consider such questions as: Is it a recent phenomenon? When did it become institutionalized and regulated? What legal and/or social issues have received special attention?
46. Consult an encyclopedia, almanac, or other suitable reference source for the total amount of consumer debt during the last 25 years. Prepare a bar graph to display the data. Are there any trends or special features to the graph?
47. Research the effect of inflation and deflation on those who owe money. Has this ever had important political ramifications?
48. What are usury laws? Are such laws in force in your state? Were they ever?

6.3 AMORTIZED LOANS

INITIAL PROBLEM

You need to borrow \$85,000 to buy a house. Loan rates are at 7%. How much will your monthly payment be for a 15-year home loan? What would the payment be for a 30-year loan?

The majority of loans involve regular payments over a period of several, or many, years. In the real world of finance, there may be factors involved that cause the APR to be different from an advertised rate. In this section we will look only at time payment loans in which rates accurately reflect the APR over the entire life of the loan.

Terminology

The word *amortize* comes from the Latin *admortiz,* which means "to bring to death." In the context of loans, it is the debt that is brought to death, not the debtor. An **amortized loan** is a simple interest loan with equal periodic payments over the length of the loan. Although the period for payments could be years or quarters as well as months, our discussion will only consider loans with monthly payments.

Each payment includes the interest that is due since the last payment and an amount to reduce the balance owed. The size of the equal payments is chosen so that once all the payments are made, the balance will be zero. That is, the loan will have been paid off. Because of rounding, the last payment may be slightly more, or less, than the other payments.

The important factors related to an amortized loan are the principal (the amount borrowed), the interest rate (the APR), the length of the loan (also called the **term of loan**), and the **monthly payment**. These four factors are interrelated. If you know any three of them, the fourth can be found. However, it is easier to find some of the factors than it is to find the others.

Charting the History of a Loan

To understand amortized loans, we will examine one over its entire history. For convenience we do this with a loan having only a few monthly payments. In reality, though, amortized loans are most often made for a period of a few years (for motor vehicles) and up to 30 years (for homes).

EXAMPLE 6.14 Chart the history of an amortized loan of \$1000 for three months at 12% with monthly payments of \$340.

Solution

When the first payment is made, $\frac{1}{12}$ of a year has passed; so the interest is $\$1000 \times .12 \times \frac{1}{12} = \$10$. The payment first goes toward paying the interest, then toward reducing the balance. This means the payment toward the balance is reduced to $\$340 - \$10 = \$330$. The \$330 is the **net payment** or **payment to principal**. The new balance for the loan after one payment is

$$\$1000 - 330 = \$670.$$

Although we could have added the interest to the balance and then subtracted the full payment, it is more convenient to find the net payment and then reduce the balance. The same steps are repeated two more times. When the second payment is made the interest charged for the time between the first and second payments is

$$\$670 \times .12 \times \frac{1}{12} = \$6.70.$$

The net payment will be \$333.30, leaving a new balance after the second payment of \$336.70.

$$\text{Net payment} = \$340 - 6.70 = \$333.30;$$
$$\text{New balance} = \$670 - 333.30 = \$336.70.$$

When the third payment is made the interest charged for the time between the second and third payments is

$$\$336.70 \times .12 \times \frac{1}{12} = \$3.37.$$

The third (and last) payment has to cover both the remaining balance and the interest due since the last payment. Therefore, the last payment must be \$340.07.

$$\text{Remaining balance} + \text{Interest due} = \$336.70 + \$3.37 = \$340.07$$

The record can be kept neatly in tabular form (Table 6.3).

TABLE 6.3

		Beginning Balance 1000.00	
Payment	*Interest*	*Net Payment*	*New Balance*
340.00	10.00	330.00	670.00
340.00	6.70	333.30	336.70
340.07	3.37	336.70	0.00

Once you know the principal, interest rate, and payment size you can chart the history of any amortized loan. Of course, this procedure will probably be stretched over a period of years. The main pitfall faced is making an error in rounding off. If you borrow from a financial institution they will do the computation for you, but you may wish to check their work. If you have an amortized loan made between private parties, such as an owner-financed home purchase, it is essential that at least one party knows how to do the calculation or has access to appropriate tables of values.

Finding the Monthly Payment

Perhaps the most important problem in dealing with amortized loans is finding the payment when you are given the loan amount, the interest rate, and the length of the loan. There are three ways to do this: (a) by using a financial or business calculator, (b) by looking it up in a table, and (c) by applying a formula and using a scientific calculator.

Using a Financial Calculator

Perhaps the easiest way to find a monthly payment is to own a financial calculator. These have built-in functions to calculate the payment required for any amortized loan. Real estate agents routinely use such calculators. The following example shows the keystrokes used on one such calculator.

EXAMPLE 6.15 Using a financial calculator, find the monthly payment on a five-year loan of $10,000 at 10% interest.

Solution

10000 [PV] (This enters the present value of $10,000.)

0.1 [÷ 12] [i] (This enters the interest rate divided by 12.)

5 [× 12] (This enters 5 × 12 = 60 months.)

[COMP] [PMT] [212.47045] (This computes the payment: $212.47.)

Using an Amortization Table

Another common way to find the payment required for an amortized loan is to consult an **amortization table**. You can look up the payment required for a variety of typical loans. Such amortization tables can be found in most business stationery stores, are relatively inexpensive, and last a lifetime. Typically, each page is devoted to one interest rate. The rows correspond to the amount of the loan and the columns correspond to the length of the loan. The entry in a particular row and column is the monthly payment required for the loan. Table 6.4 shows part of such a table; a more complete table is provided in the Appendix.

TABLE 6.4
AMORTIZATION TABLE AT 10%

Amount	5 Years	10 Years	15 Years	20 Years	25 Years	30 Years	35 Years	40 Years
100	2.13	1.33	1.08	.97	.91	.88	.86	.85
200	4.25	2.65	2.15	1.94	1.82	1.76	1.72	1.70
500	10.63	6.61	5.38	4.83	4.55	4.39	4.30	4.25
1000	21.25	13.22	10.75	9.66	9.09	8.78	8.60	8.50
2000	42.50	26.44	21.50	19.31	18.18	17.56	17.20	16.99
5000	106.24	66.08	53.74	48.26	45.44	43.88	42.99	42.46
10000	212.48	132.16	107.47	96.51	90.88	87.76	85.97	84.92
20000	424.95	264.31	214.93	193.01	181.75	175.52	171.94	169.83
50000	1062.36	660.76	537.31	482.52	454.36	438.79	429.84	424.58
100000	2124.71	1321.51	1074.61	965.03	908.71	877.58	859.68	849.15

Example 6.16 Find the monthly payment on a five-year loan of $10,000 at 10% interest.

Solution

We look for the entry in the row for the amount of 10000 and the column for a term of five years. The row and column have been shaded in Table 6.4. We find the monthly payment is $212.48. This compares to the value of $212.47 found in Example 6.15. The difference is due to rounding.

A second type of amortization table assumes a standard loan size such as $1000. The rows then correspond to the interest rate, and the columns correspond to the length of the loan. The entry in the table is the payment per $1000 borrowed. To get the correct payment for a particular loan, multiply the entry by the number of thousands of dollars borrowed. The result must be rounded up to get the proper payment. If you were to round down, every payment would be somewhat smaller, and the loan would not be paid off in the assigned number of payments. When you round up, the final payment will be less than the regular payment.

EXAMPLE 6.17 Find the monthly payment on a 10-year loan of $13,000 at 12% interest.

Solution

We look for the entry in the row for a 12% interest rate and the column for a 10-year loan. The row and column have been shaded in Table 6.5. We find $14.347095. Since this is the payment per $1000, we multiply by 13 (for a $13,000 loan) to get

$$\$14.347095 \times 13 = \$186.512235.$$

The correct payment is $186.52.

TABLE 6.5

AMORTIZATION TABLE FOR $1000 LOAN

Percent	5 Years	10 Years	15 Years	20 Years	25 Years	30 Years
5	18.871234	10.606552	7.907936	6.599557	5.845900	5.368216
6	19.332802	11.102050	8.438568	7.164311	6.443014	5.995505
7	19.801199	11.610848	8.988283	7.752989	7.067792	6.653025
8	20.276394	12.132759	9.556521	8.364401	7.718162	7.337646
9	20.758355	12.667577	10.142666	8.997260	8.391964	8.046226
10	21.247045	13.215074	10.746051	9.650216	9.087007	8.775716
11	21.742423	13.775001	11.365969	10.321884	9.801131	9.523234
12	22.244448	14.347095	12.001681	11.010861	10.532241	10.286126
13	22.753073	14.931074	12.652422	11.715757	11.278353	11.061995
14	23.268251	15.526644	13.317414	12.435208	12.037610	11.848718
15	23.789930	16.133496	13.995871	13.167896	12.808306	12.644440

Using a Formula

If one has access to a scientific calculator with a [x^y] key, the following formula can be used to find a monthly payment.

> ***MONTHLY PAYMENT FORMULA***
>
> If P is the amount of the loan, r is the annual percentage rate as a decimal, and t is the length of the loan in years, then the monthly payment, PMT, is given by
>
> $$PMT = \frac{P \times \frac{r}{12} \times (1 + \frac{r}{12})^{12t}}{[(1 + \frac{r}{12})^{12t} - 1]}.$$

Clearly, you do not use this formula without a calculator, and it must be a calculator with an [x^y] key or the equivalent.

EXAMPLE 6.18 Find the monthly payment for a \$1000 loan for three years at 10% using the formula.

Solution

Here $P = \$1000$, $r = 0.10$, and $t = 3$. By the formula then

$$\begin{aligned} PMT &= \frac{\$1000 \times (\frac{0.10}{12}) \times (1 + (\frac{0.10}{12}))^{12\times 3}}{[(1 + (\frac{0.10}{12}))^{12\times 3} - 1]} \\ &= \frac{\$1000 \times 0.008333 \times 1.008333^{36}}{[(1.008333)^{36} - 1]} \\ &\approx \$32.27. \end{aligned}$$

These calculations can be done with one long string of keystrokes on a calculator using parentheses. However, due to the complexity of the formula and the fact that $1 + (\frac{0.10}{12})^{12\times 3}$ appears twice, a calculator memory key is useful.

Step 1: Clear your calculator memory. Then calculate $\left(1 + (\frac{0.10}{12})\right)^{12\times 3}$ and store it in memory.

(1 [+] 0.1 [÷] 12) [x^y] 36 [=] [1.348182] [STO]

This stores the value in memory, and you can recall it as you need it.

Step 2:

1000 [×] 0.1 [÷] 12 [×] [RCL] [÷] ([RCL] [−] 1) [=] [32.267188]

Thus a payment of \$32.27 per month is needed to pay off a \$1000 loan at 10% in three years. ∎

INITIAL PROBLEM SOLUTION

You need to borrow \$85,000 to buy a house. Loan rates are at 7%. How much will your monthly payment be for a 15-year home loan? What would the payment be for a 30-year loan?

Solution

We are given an interest rate of 7%. Typically, mortgages are for a term of 15 or 30 years. For a 15-year loan at 7%, Table 6.5 gives a payment of \$8.988283 per \$1000. The payment on an \$85,000 loan would be

$$\$8.988283 \times 85 \approx \$764.01.$$

For a 30-year loan at 7%, the table gives a payment of \$6.653025 per \$1000. The payment on an \$85,000 loan would be

$$\$6.653025 \times 85 \approx \$565.51.$$

The choices are \$764.01 per month for 15 years or \$565.51 per month for 30 years. These payments cover the principal and interest only. The typical house payment will also include amounts for taxes and insurance.

Problem Set 6.3

1. A \$2000 loan is made at 10% interest with monthly payments of \$42.50 for five years. What is the balance after the first payment is made?
2. A loan of \$5000 is made at 8% interest with monthly payments of \$101.40 for five years. What is the balance after the first payment is made?
3. A new car is purchased with a trade-in and a 4.9% loan on the balance of \$13,000. If the payments are \$298.79 a month for 48 months, what is the balance on the loan after the first payment is made?
4. Tom has to have his car painted. He finances the bill of \$1350 with a 12.9% loan for two years from the credit union. If his monthly payments are \$64.12, what is the balance on the loan after the first payment is made?
5. Chart the first three months' history of an amortized loan of \$5000 for five years at 12% with monthly payments of \$111.23.
6. Chart the first three months' history of an amortized loan of \$10,500 for four years at 12% with monthly payments of \$276.51.
7. Chart the history of a loan of \$600 that is paid back with five monthly payments of \$123.78 based on a 12.5% rate. What is the amount of the final payment?
8. Chart the history of a loan of \$900 that is paid back with four monthly payments of \$231.25. If the rate is 13.25%, what is the amount of the final payment?
9. Use Table 6.4 to find the monthly payment on a 10-year loan of \$10,000 at 10% interest.
10. Use Table 6.4 to find the monthly payment on a 20-year loan of \$50,000 at 10% interest.
11. Use Table 6.4 to find the monthly payment on a five-year loan of \$18,000 at 10% interest.
12. Use Table 6.4 to find the monthly payment on a 15-year loan of \$28,000 at 10% interest.
13. Use Table 6.5 to find the monthly payment on a five-year loan of \$18,000 at 12% interest.
14. Use Table 6.5 to find the monthly payment on a 10-year loan of \$15,000 at 9% interest.
15. Use Table 6.5 to find the monthly payment on a 15-year loan of \$22,500 at 6% interest.
16. Use Table 6.5 to find the monthly payment on a 20-year loan of \$25,750 at 8% interest.
17. Use the formula to find the monthly payment on a 10-year loan of \$20,000 at 10% interest. Verify the value by comparing it to Table 6.4.
18. Use the formula to find the monthly payment on a five-year loan of \$5000 at 10% interest. Verify the value by comparing it to Table 6.4.
19. Find the monthly payment on a five-year loan of \$1000 at 9% interest. Use the formula, and compare your result with the value in Table 6.5.

20. Find the monthly payment on a 10-year loan of $1000 at 12% interest. Use the formula, and compare your result with the value in Table 6.5.

21. Find the monthly payment for a $6500 loan for 30 months at 9%.

22. Find the monthly payment for a $4800 loan for two years at 9.8%.

23. What is the monthly payment on a loan of $4850 for 20 months at 7.25%?

24. What is the monthly payment on a loan of $10,800 for four years at 11.75%?

25. What is the monthly payment on a car loan for $6725 financed for four years at 12.8% interest?

26. What is the monthly payment needed to finance a new car purchase of $12,400 for 40 months at 7.9% interest?

27. What is the monthly payment on a furniture purchase of $3250 financed for two years at 10.5% interest after a 20% down payment is made?

28. A home priced at $85,000 is sold with a 10% down payment and the balance financed at 8% for 30 years. What is the monthly payment?

29. John bought a new car for $16,285 by paying 20% down and financing the balance at 11% for five years. What will his monthly payment be?

30. After graduation, Wendy took a tour to the major art museums of eastern Europe. The tour was financed with a down payment of 10% and the balance paid in 30 monthly payments at 11.5% interest. The tour package Wendy chose was priced at $2795. What was her monthly payment?

31. Use the amortization tables to find the size of loan that can be financed at 9% for 20 years with a monthly payment of $500.

32. Use the amortization tables to find the size of loan that can be financed at 12% for 15 years with a monthly payment of $300.

33. Use Table 6.5 to find the size of loan that can be financed at 9.5% for 15 years with a monthly payment of $600. Since 9.5% isn't listed on the table, use an appropriate approximation.

34. As in Problem 33, use Table 6.5 to find the size of loan that can be financed at 8.75% for 10 years with a monthly payment of $250.

35. Tim needs to buy a car. After going over his budget, he decides he can afford $250 a month for a car payment. If he pays no money down and gets financing for five years at 8% interest, how much can he afford to pay for a car?

36. The Romeros decide to get a home improvement loan. After going over their finances, they determine that the most they can budget for a monthly payment is $300. If they can get a 10-year loan at 9%, what is the maximum loan possible?

37. What is the approximate annual interest rate for a 30-year loan of $40,000 with monthly payments of $321.85?

38. What is the approximate annual interest rate for a 25-year loan of $30,000 with monthly payments of $231.55?

39. What is the approximate annual interest rate for a 15-year loan of $25,000 with monthly payments of $292.68?

40. What is the approximate annual interest rate for a 10-year loan of $50,000 with monthly payments of $626.30?

Extended Problems

41. Compare the terms available through local financial institutions for amortized loans to buy the following items.
    (a) used car **(b)** new car
    (c) manufactured home **(d)** conventional home

42. Discuss why the terms you found in Problem 41 differ from each other.

43. Typically, interest payments for a loan secured by your home are deductible on your income tax.
    (a) If you had the cash to pay for your house, should you do so, or should you finance the house and invest the cash?
    (b) How does the answer to (a) depend on interest rates and tax bracket?

The Formula for Finding the Loan Amount

If you are given the interest rate, the length of the loan, and the monthly payment, you can use the formula for finding the monthly payment "in reverse" to find the amount of the loan. If

r = interest rate (as a decimal),
t = length of the loan (in years)
PMT = the monthly payment, and
P = principal of the loan,

then

$$P = PMT \times \frac{12}{r} \times \left[1 - \frac{1}{(1 + \frac{r}{12})^{12t}}\right].$$

44. Use the formula to find the size of loan that can be financed at 9.5% for 15 years with a monthly payment of $600.

45. Use the formula to find the size of loan that can be financed at 10.75% for 20 years with a monthly payment of $420.

46. Verify the formula given. Begin with the formula for finding the monthly payment, and solve for P.

6.4 BUYING A HOUSE

INITIAL PROBLEM

Suppose you have saved $15,000 towards a down payment on a home, and your total household income is $35,000 per year. What is the most you could afford to pay for a home? Assume that (1) your insurance costs will be 0.25% of the value of your home, (2) your taxes will be 2% annually, (3) your closing costs will be about $2,000, and (4) you can obtain a fixed rate mortgage for 30 years at 8% interest.

Not only is the purchase of a new home the biggest financial commitment in most people's lives, it is also one of the most complicated. Many factors beyond the price of the home have to be considered, beginning with the ability to pay the initial costs as well as the monthly costs that may last for 30 years. This section will cover the basic mechanics of a home mortgage. We will not cover some topics of interest, such as the tax advantages of home ownership, which are beyond the scope of this book.

Affordability Guidelines

History

Part of the English feudal system involved a duty to the lord on whose land you had tenure- for example, military service. Most of the Colonies adopted this system, imposing the least burdensome duty of "socage," which was usually payment of rent to the Crown. This system was distasteful to the colonists and led to incidents of rebellion as early as 1676.

A central part of the "American Dream" is owning your own home. This requires money, and most people must borrow the bulk of that money. We have already covered the general topic of loans, but when real estate is involved there are a number of additional conditions that can make such transactions extremely complex.

A natural question might be "Why is real estate so complicated?" One answer is "History!" The laws governing real estate in most of the United States evolved from the laws of England while the states were still colonies. Those English real estate laws were mainly derived from ancient feudal laws and from even earlier common law predating the Norman conquest. Unfortunately, as real estate law evolved, it was not simply a case of the new replacing the old. Instead, the new was added to the old, making a progressively more complex legal structure. One consequence is that there is no single document to prove conclusively that a particular person owns a particular property. In any case, we cannot cover real estate law, nor would we want to. We will focus on the main mathematical issues that affect the ordinary person. If you purchase a home, we trust that you will have professionals help sort out the legal details.

Traditionally, a few general guidelines have been used to estimate how much a buyer could afford to spend on a house. Here are two of the most common ones.

1. The home you purchase should not cost more than three times your annual family income, assuming a standard down payment of 20%.
2. You should limit your monthly housing expenses, including mortgage payment, property taxes, and homeowner's insurance, to no more than 25% of your monthly gross income (that is, income before deductions).

If your planned house purchase fits under both of these guidelines, then you can almost surely afford it.

EXAMPLE 6.19 If your annual family income is \$30,000, what do the traditional affordability guidelines tell you regarding a purchase price and monthly expenses for your potential home?

Solution

The houses you consider should not cost more than

$$3 \times \$30{,}000 = \$90{,}000$$

and the monthly expenses for mortgage payment, property taxes, and homeowner's insurance should not exceed

$$0.25 \times \frac{1}{12} \times \$30{,}000 = \$625.$$

In practice, the most important question is whether or not a bank or other financial institution will approve your loan application. Among the things that will be considered is your other debt. Having car payments and credit card balances may affect your ability to buy a house. As of this writing, many banks are allowing up to 38% of the borrower's monthly income to go for mortgage payments, property taxes, and homeowner's insurance. For the situation in the Example 6.19, assuming there are no other significant debts, the limit on monthly housing expenses may go as high as

$$0.38 \times \frac{1}{12} \times \$30{,}000 = \$950.$$

EXAMPLE 6.20 Suppose Andrew and Barbara both have jobs, each earning \$24,000 per year, and they have no debts. What are the low and high estimates of how much they can afford to pay for monthly housing expenses?

Solution

The low estimate for acceptable monthly housing expenses is 25% of gross monthly income, or in the case of Andrew and Barbara

$$0.25 \times \frac{24{,}000 + 24{,}000}{12} = \$1000.$$

The high estimate for acceptable monthly housing expenses is 38% of gross monthly income, or

$$0.38 \times \frac{24{,}000 + 24{,}000}{12} = \$1520.$$

The Mortgage

A **mortgage** is a loan that is guaranteed by real estate. If the borrower fails to make the payments, the lender can take control of the property. Technically only certain loans secured by real estate are called mortgages, but we will use the term in the sense of everyday conversation rather than its full legal meaning.

There are many types of mortgages available. In an era of financial stability, such diversity would not make much sense, but the 1970s and 1980s were decades of high inflation and high interest rates. Those factors forced financial institutions to introduce new types of mortgages to meet the needs of clients and attract otherwise marginal borrowers.

The two main categories of mortgage are *fixed rate* and *adjustable rate*. For a **fixed rate mortgage**, the interest rate is set, once and for all, at the time the loan is made. For an **adjustable rate mortgage**, the interest rate can change from year to year. The actual interest rate is usually a specified amount higher than some particular financial index (for example, the interest rate paid by Treasury bonds); often there is also a limit (called a **cap**) on how much the interest rate is allowed to rise in a single year.

The second major distinction among mortgages is the **term** of the mortgage. Typically the choices are 15 year or 30 year. The longer term loan usually carries a higher interest rate because the money is used for a longer period of time, and the lender's risk is extended.

Another variable in choosing a mortgage loan is commonly referred to as **points**. One point is one percent of the amount of the loan. Points are generally charged in two ways: (1) a **loan origination fee** for making the loan at all and (2) a **discount charge** for offering a lower interest rate. For a fixed rate mortgage the combined effect of the interest rate, loan origination fee, and discount charge can be summed up in the annual percentage rate. For an adjustable rate mortgage the fee and discount charge are typically smaller, and the annual percentage rate cannot be computed because the interest rate will be changing.

EXAMPLE 6.21 If you are going to borrow \$80,000 for a home and there is a one-point loan origination fee and a one-point discount charge, what will be your added costs?

Solution

Each of these expenses will cost 1% of \$80,000, or \$800. Thus your added costs will be \$1600.

One question is "Do I want a fixed rate mortgage or an adjustable rate mortgage?" Unfortunately, the answer depends on what is going to happen to interest rates in the future. For example, if you knew that interest rates were going to go down significantly over the life of your mortgage, then you would want to have an adjustable rate mortgage so that the interest rate you pay would go down also. On the other hand, if you knew that interest rates were going to go up over the life of your mortgage, then you would want a fixed rate mortgage to lock in the initial lower interest rate. Of course, if you really knew in advance what interest rates were going to do, you could use your psychic powers to make significant amounts of money in the stock market, since changes in interest rates have well-known effects on stocks and bonds.

In the real world, you do not have advance knowledge of the direction of interest rates, although the rates are carefully monitored and analyzed by financial professionals. Interest rates reflect national economic policy and economic conditions, with the Federal Reserve setting the pace. You must make your choice on the basis of what you can afford when you initially borrow the money, your expectations concerning your future income, and the amount of uncertainty you can accept. If rates are changing, timing may be the most important factor.

Table 6.6 provides a sample of mortgage loan information.

TABLE 6.6
INTEREST RATE TABLE

Banks	Term	Rate	F + DC	APR
American Bank	15	6.625	1.75	6.94
	30	7.125	1.75	7.32
Broward Co. Bank	15	6.375	1.775	6.706
	30	6.875	1.925	7.097
First Interstate	15	6.5	1.75	6.865
	30	7	1.75	7.23
First Security Bank	15	6.5	2	6.9199
	30	7	2	7.2672
Key Largo Bank	15	6.5	2.125	6.842
	30	7	2.125	7.2142
Liberty Federal Bank	15	6.625	1	6.818
	30	7.125	1	7.246
U.S. Bank	15	6.375	2.25	6.777
	30	7	1.875	7.215
West One Bank	15	6.5	1.75	6.7892
	30	7	1.75	7.182

Similar tables are published weekly in most major newspapers. (Most of the columns in the tables are self-explanatory, but the "F + DC" column is not. This column gives the points charged.) Scanning through the table, you can see that the annual percentage rate on a 30-year fixed mortgage was about 7%.

The Down Payment

A typical **down payment** on a house is 20% of the total value of the house. The published rates in Table 6.6 assume such a 20% down payment. If the down payment is 20% of the value of the property, then the value of the property you can buy is the down payment divided by 0.20.

EXAMPLE 6.22 If you have \$20,000 for a down payment, what is the highest priced home you can afford if a 20% down payment is required?

Solution

$\frac{\$20{,}000}{0.20} = \$100{,}000$. Thus you can shop for a \$100,000 house.

In general

$$\text{total value} = \text{down payment dollars} \div \text{down payment percentage}$$

where the percentage must be converted to a decimal.

A lender prefers a large percentage down payment as protection against the borrower defaulting on the loan. If payments are missed and the lender has to take control of the property, the down payment is a significant part of the value that can be recovered. This lessens the risk to the lender. If a smaller down payment is made, you must expect to pay in some way for the lender's increased risk. The interest rate may be higher, or you may be required to purchase Private Mortgage Insurance—which has a price in points—to insure the lender against a default on the loan. Another possibility for qualified buyers is an FHA (Federal Housing Administration) loan, which not only insures the lender against default, but also requires a lower down payment. Down payments smaller than 20% are becoming more common, in particular in programs aimed at attracting first-time buyers. Returning to Example 6.22, if the \$20,000 is used to make a 5% down payment, then the value of the property being considered could be as high as

$$\frac{\$20{,}000}{0.05} = \$400{,}000.$$

Note that if you bought the \$400,000 house with 5%, or \$20,000, down, that would mean a loan of \$380,000. Even with a low rate of 6%, the interest alone would be \$1900 the first month!

EXAMPLE 6.23 The Garcia family purchases a home for \$150,000. They pay 20% down and pay all the closing costs. If the balance is financed at 7% for 30 years, what will their monthly payment be for principal and interest?

Solution

The 20% down payment is $\$150{,}000 \times 0.20 = \$30{,}000$. The balance to finance is $\$150{,}000 - \$30{,}000 = \$120{,}000$. Using Table 6.5, we see that the payment for principal and interest on \$120,000 at 7% over 30 years is

$$120 \times 6.653025 \approx \$798.37.$$

Tidbit

Because of the complex way ownership of real estate is determined, it is necessary to purchase insurance against the possibility that someone else actually owns the house you bought and paid for! This is called title insurance.

Remember the Points

The loan origination fee and the discount charge were mentioned earlier. Table 6.6 shows that the F + DC factor varies from a low of 1 point to a high of 2.25 points. If the down payment is less than 20%, there may well be mortgage insurance points. Additionally, there may be other charges when the transaction is finalized. This finalizing of the purchase is called the **closing**, and the additional expenses are called **closing costs**. If you purchase a home, you should ask your realtor in advance how much this will cost, so that you can have enough in your checking account to cover it. The costs vary from state to state, and whether the buyer or seller pays for a particular item may be negotiable.

INITIAL PROBLEM SOLUTION

Suppose you have saved \$15,000 toward a down payment on a home, and your total household income is \$35,000 per year. What is the most you could afford to pay for a home? Assume that (1) your insurance costs will be 0.25% of the value of your home, (2) your taxes will be 2% annually, (3) your closing costs will be about \$2000, and (4) you can obtain a fixed rate mortgage for 30 years at 8% interest.

Solution

First, we estimate the maximum loan you can afford. In addition to the down payment, a limiting factor to the amount of loan you can obtain is the monthly payment you can afford to make. Your monthly obligation includes your payment, home insurance, and any real estate taxes (which will vary from state to state).

Your total income is \$35,000 per year,and you have \$15,000 available to use for the purchase. However, \$2000 will be needed for closing costs, leaving \$13,000 for your down payment. By the first affordability guideline, you might look for a home in the $3 \times \$35{,}000 = \$105{,}000$ range. By the second affordability guideline, you should be able to afford to make monthly payments of 25% of $\frac{1}{12}$ of \$35,000, or about \$730 per month. The following calculations will determine if you can afford a house in this price range.

Cost of the house	\$105,000
Down payment	13,000
Amount needed to finance	92,000
Monthly payment on 30 year 8% loan (use Table 6.5)	675
Cost of insurance and real estate taxes ($\frac{1}{12}$ of 2.25% of \$105,000)	197
Total monthly payment	\$872

According to the second affordability guidelines, you will be able to afford a monthly payment of \$730. Since \$872 exceeds \$730, there are several things you might do to afford the house in this price range:

1. Wait for interest rates to fall.
2. Increase your income(s).
3. Come up with a larger down payment.

With the same down payment, a \$100,000 house with a 7% loan would require a monthly payment of \$767—still a bit too high. If you increased the down payment to \$25,000, then the monthly payment for the 8% loan would be \$738, which is close enough to have a loan approved.

If the bank uses the more liberal guideline of 38%, you could afford a monthly payment of approximately \$1110, and the loan would most likely be approved.

Problem Set 6.4

1. If a family's annual income is \$35,000, what do the two affordability guidelines tell them regarding the purchase price and monthly payments for a potential home purchase?
2. If a family's annual income is \$52,800, what do the two affordability guidelines tell them regarding the purchase price and monthly payments for a potential home purchase?
3. Martha and Ed have a combined income of \$43,550 a year. Based on the two affordability guidelines, what are their considerations concerning price and monthly payments if they decide to buy a house?
4. After he graduates from college, Philip is going to invest in a house rather than rent. He anticipates going to work for a software engineering company for at least \$32,000 per year. If he follows the affordability guidelines, what is the maximum he should consider paying for a house? What about monthly housing expenses?
5. Using the two affordability guidelines, what is the maximum Paul and Ann should pay for a house if his annual income is \$22,000 and hers is \$31,000? How much should they be willing to pay for monthly housing expenses?
6. Richard and Rose Sheng are considering buying a new home. He has an annual income of \$35,000 and hers is \$24,000. What do the affordability guidelines tell them regarding purchase price and monthly housing expenses?
7. If a couple has a combined monthly income of \$3650, what are the low and high estimates of what they can afford to pay for monthly housing expenses?

8. If a couple has a combined monthly income of $2980, what are the low and high estimates of what they can afford to pay for monthly housing expenses?

9. The Baileys are going to borrow $95,000 for a new home. If there is a one-point loan origination fee and a one-point discount charge, what are their costs for these two items?

10. If there is a one-point loan origination fee and a one-point discount charge on a home loan of $78,500, what is the total for these two charges?

11. A home loan of $92,000 is subject to a 1.5-point loan origination fee and a 1-point discount charge. What is the total for these two charges?

12. There is a 1.25-point loan origination fee and a 1-point discount charge on a new home loan for $165,000. What is the total for these two charges?

13. John and Sue have been saving their money to buy a home and have $23,000 available for a down payment. The realtor says that they should expect to need 20% for a down payment. What is the maximum price they will be able to afford based upon the down payment?

14. Referring to Problem 13, if John and Sue are able to qualify for a special loan program for new home buyers, the realtor tells them they will only need a down payment of 10%. If this is true, how much will they be able to afford based upon their down payment?

15. If a down payment of 20% is needed and a couple has $18,000 for a down payment, what price home could they buy? What annual income should they have in order to follow the affordability guideline regarding purchase price?

16. If a 10% down payment is needed for a home in a new subdivision, what is the maximum price the Addams can consider if they have $15,000 for a down payment?

17. The Davis family sold some property and are in the market for a new home. They have $18,000 available for the home purchase and initial costs. What is the maximum price they will be able to pay if they need $2500 for closing costs and 10% for a down payment?

18. Referring to Problem 17, suppose the Davis family finds a house they wish to buy priced at $135,000, but the seller requires a 20% down payment. How much additional money does the Davis family need?

19. Use Table 6.5 to find the monthly payment for principal and interest on a loan of $72,000 at 12% interest for 30 years.

20. Use Table 6.5 to find the monthly payment for principal and interest on a $91,800 loan at 10% interest for 15 years.

21. A family decides to buy a home at an agreed upon price of $135,000. After making a down payment of 20% and paying all closing costs, the balance was financed at 8% for 30 years. What is the monthly payment for principal and interest? Use Table 6.5.

22. A home was sold for $92,500 with a down payment of 20%. The remaining balance was financed at 9% for 20 years. What is the monthly payment for principal and interest? Use Table 6.5.

23. A home priced at $127,500 sold with a 10% down payment. The remaining balance was then financed at 9.5% for 30 years. Estimate the monthly payment for principal and interest. Use Table 6.5 and average the amounts corresponding to 9% and 10%.

24. A home sold for $115,500 with a down payment of 10%. The remaining balance was financed at 10.5% for 20 years. Estimate the monthly payment for principal and interest. Use Table 6.5 and average the amounts corresponding to 10% and 11%.

25. A home was sold for $95,500 with a down payment of 10%. The remaining balance was financed at 9.6% for 30 years. Find the monthly payment for principal and interest using the monthly payment formula.

26. A loan balance of $74,200 was financed at 8.85% for 20 years. Find the monthly payment for principal and interest using the monthly payment formula.

Problems 27 through 30

Many banks and other lenders often require extra funds be collected to cover yearly payments for property taxes and insurance. These additional funds are generally included as part of the monthly mortgage payment and placed in a "reserve" account until they are due. Use Table 6.5, together with the information on taxes and insurance, to find the complete monthly payment including principal, interest, taxes, and insurance.

27. Assessed value = $150,000
Loan amount = $115,000
Rate = 9%
Term = 30 years
Taxes = 2.5% of assessed value
Insurance = $650 per year

28. Assessed value = $120,000
Loan amount = $105,000
Rate = 8%
Term = 20 years
Taxes = 2.5% of assessed value
Insurance = $480 per year

29. Assessed value = $127,700
Loan amount = $75,000
Rate = 10%
Term = 15 years
Taxes = 2.25% of assessed value
Insurance = $740 per year

30. Assessed value = $225,000
Loan amount = $165,000
Rate = 9%
Term = 30 years
Taxes = 1.85% of assessed value
Insurance = $1260 per year

Extended Problems

31. Find current data on interest rates for mortgages in your area.

32. Consider the difference between 15-year and 30-year mortgages on a $100,000 home. Use typical current rates in your area if possible.
    (a) How much is the difference in monthly payments?
    (b) How much is the difference in the total money paid over the entire loan? Include points if that information is available.
    (c) Which seems to be the better deal?

33. The saying "Possession is nine points of the law" is due to the 17th-century writer Thomas Fuller. The fact that ownership of real estate is proved by title refutes Fuller for that one form of property. For what other types of property is ownership provable by title?

34. The familiar frame house was a technical innovation developed in the 1830s in Chicago, Illinois, and Rochester, New York. Why was this type of construction invented, and what was its importance?

35. What was the significance of Levittown to widespread home ownership?

36. What programs are available through the federal government that promote home ownership through low down payments or low interest rates?

6.5 ANNUITIES AND SINKING FUNDS

INITIAL PROBLEM

In 18 years you would like to have $50,000 saved for your child's college education. At 6% interest, compounded monthly, what monthly deposit must be made to accomplish this goal?

Many of our important purchases or financial commitments are taken care of after the fact through loans and mortgages. However, this is not a practical strategy for many situations, such as retirement. Many people also prefer to have their funds set aside before obligating the funds, while government entities may legally be required to do so, and businesses may find it necessary or prudent. These situations require thoughtful planning, and the key to success is making appropriate regular payments to carry out the plan.

Terminology

> **History**
>
> *Large amounts of cash sitting in a sinking fund have sometimes been an irresistible temptation for governments and companies and have been raided for other purposes than originally intended.*

An **annuity** is an account into which a sequence of equal, regular payments are made. For example, an account to which a person contributes one hundred dollars a month until retirement is an annuity. An annuity is really just a saving strategy. It is a useful strategy for two reasons. First, the regular nature of the deposits helps the saver maintain discipline. Second, if the payments are made over a long time period, the interest received contributes significantly to the growth of the annuity, especially the compounding of the interest.

Unfortunately, the financial world also has another meaning for the word *annuity.* In the second context, if you purchase an annuity you will receive a sequence of equal, regular payments. To distinguish between the two meanings, we will call the latter type of annuity an **income annuity**.

A **sinking fund** is a type of annuity in which the goal is to have a particular amount of money saved at the end of a given time period. Historically, the term *sinking fund* referred to the method state and local governments used to accumulate enough money to pay off bonds issued for improvements such as roads and bridges. To raise money for such improvements, bonds (a type of promissory note) would be sold at a discount from the face value (the value stated on the bond). Bonds are generally issued for periods of 10 to 30 years, and the issuer of the bonds has no obligation to pay until the end of the bonding period, known as the maturity date. When the maturity date arrives, the issuer pays the holders of the bonds the face value of the bonds. This is called "redeeming" the bonds. To ensure that the money for redemption was available when the maturity date arrived, regular deposits would be made to a sinking fund during the life of the bonds. The size of the regular payments was chosen so that there would be sufficient money in the fund.

There are two basic questions with annuities. One is to determine how much will accumulate over time when given periodic payments are made, and the other is to determine what periodic payments will be necessary in order to accumulate a specific amount in a given period of time. In a sense, these questions are the reverse of those when amortizing loans.

Suppose you want to pay for something that costs more than you can afford from your paycheck, say a vacation. One strategy is to borrow the money; another is to set aside a smaller amount from each paycheck until you have accumulated enough for what you want to do. If you regularly set aside a given amount of money in an account that pays interest, then you have an annuity.

EXAMPLE 6.24 Mary wants to take a nice vacation trip, so she begins setting aside \$250 per month. If she deposits this money on the first of each month in a savings account that pays 6% interest compounded monthly, how much will she have at the end of 10 months?

Solution

Mary's first payment of \$250 will earn interest for 10 months. In order to make the calculation, we use the compound interest formula

$$F = P\left(1 + \frac{r}{m}\right)^{mt}, \text{ where } r = 0.06,\ m = 12, \text{ and } t = \frac{10}{12} = \frac{5}{6} \text{ (year)}.$$

Ten months later that \$250 will have grown to

$$250 \times \left(1 + \frac{0.06}{12}\right)^{12(5/6)} = 250 \times (1.005)^{10} = 262.79.$$

The amount of \$262.79 is referred to as the future value of the first payment. The future value of the second payment is

$$250 \times (1.005)^9 = 261.48,$$

since that payment earns interest over nine months. The future values of the other payments are as follows:

Payment	Future Value	Payment	Future Value
3rd	$250 \times (1.005)^8 = 260.18$	4th	$250 \times (1.005)^7 = 258.88$
5th	$250 \times (1.005)^6 = 257.59$	6th	$250 \times (1.005)^5 = 256.31$
7th	$250 \times (1.005)^4 = 255.04$	8th	$250 \times (1.005)^3 = 253.77$
9th	$250 \times (1.005)^2 = 252.51$	10th	$250 \times (1.005)^1 = 251.25$

Totaling up the future values, we see that Mary will be able to withdraw \$2569.80 for her vacation trip. Since she only deposited \$2500, the additional \$69.80 was interest earned.

Notice that the preceding calculation used the fact that the compounding period for the interest is the same as the length of time between payments, called the **payment period**. An annuity for which the interest compounding period is the same as the payment period is called a **simple annuity**. We will restrict our attention to simple annuities. The term of an annuity is the length of time from the beginning of the first payment period until the end of the last payment period. In Example 6.24, the term of the annuity was 10 months. At the end of the term of the annuity, the annuity is said to have **expired** and the money may be withdrawn.

Payments to an annuity may be due at the beginning of each payment period, as in the example, or at the end of each payment period. An annuity for which payments are due at the beginning of each payment period is called an **annuity due** (Example 6.24 illustrates an annuity due). An annuity for which payments are due at the end of each payment period is called an **ordinary annuity**. The difference between an annuity due and an ordinary annuity is that each payment to an ordinary annuity will be on deposit for one less payment period. Therefore, every payment to an ordinary annuity will earn interest for one less compounding period than an annuity due. The difference can be significant.

Finding the Future Value of An Annuity

If an annuity involves many payments, then working out the future value of each payment and adding them together as we did in the example will be too time-consuming. Instead, the following formulas can be used to find the total at the end of the term. The amount of money available for withdrawal when the annuity expires is called the **future value of the annuity** and will be represented by the letter F in the various formulas in this section. We will also use i to represent the **periodic interest rate**, which is the annual interest rate, r, divided by the number of periods.

FUTURE VALUE OF AN ANNUITY

If PMT is the payment size, r is the annual rate (expressed as a decimal), and m is the number of payment periods per year; then the total number of payment periods is $n = mt$, the periodic rate is $i = \frac{r}{m}$, and the future value is

$$F = PMT \times \frac{(1 + \frac{r}{m})^{mt} - 1}{\frac{r}{m}} = PMT \times \frac{(1 + i)^n - 1}{i}$$

for an ordinary annuity, and

$$F = PMT \times (1 + \tfrac{r}{m}) \times \frac{(1 + \frac{r}{m})^{mt} - 1}{\frac{r}{m}}$$

$$= PMT \times (1 + i) \times \frac{(1 + i)^n - 1}{i}$$

for an annuity due.

It is when n, the total number of payment periods, is large that the effect of compound interest becomes important.

EXAMPLE 6.25 Find the future value of an ordinary annuity with a term of 25 years, payment period of one month, payment size of \$50, and annual interest rate of 6%.

Solution

With monthly payments, we have $n = 12 \times 25 = 300$. The interest rate per payment period is $i = 0.06 \div 12 = 0.005$. The payment size is $PMT = \$50$. Using a calculator to apply the ordinary annuity formula, the future value of the annuity is

50 [×] [[ ] [[ ] 1 [+] .005 [ ]] [x^y] 300 [−] 1 [ ]] [÷] .005 = [34649.69812],

or $F = \$34{,}649.70$.

Note that in this example the total of 300 \$50 payments is only \$15,000. This means that the interest earned is over \$19,000. In many annuities, particularly retirement accounts, the interest earned will significantly exceed the amount contributed by the saver.

Finding Payments for Sinking Funds

In setting up the formulas for the future value of an annuity, we assumed that the payment size had been chosen in advance and that the future value was to be found. The distinguishing feature of a *sinking fund* is that the future value is chosen in advance, and the payment size has to be found. We will continue to use PMT for the payment size, i for the interest rate per period, and n for the total number of periods. The formulas for sinking funds are as follows:

PAYMENTS FOR SINKING FUNDS

If PMT is the payment size, F is the future value, r is the annual rate (expressed as a decimal), and m is the number of payment periods per year, then the total number of payment periods is $n = mt$, the periodic rate is $i = \frac{r}{m}$, and the payment size is

$$PMT = F \times \frac{\frac{r}{m}}{(1 + \frac{r}{m})^{mt} - 1} = F \times \frac{i}{(1+i)^n - 1}$$

for an ordinary annuity, and

$$PMT = F \times \frac{\frac{r}{m}}{(1 + \frac{r}{m}) \times [(1 + \frac{r}{m})^{mt} - 1]}$$

$$= F \times \frac{i}{(1+i) \times [(1+i)^n - 1]}$$

for an annuity due.

EXAMPLE 6.26 Suppose you decide to use a sinking fund to save $10,000 for a car. If you plan to make 60 monthly payments and you receive 1% interest per month, what is the required payment for an ordinary annuity?

Solution

We apply the formula for payments for sinking funds with $F = \$10{,}000$, $n = 60$, and $i = .01$. We find

$$PMT = 10{,}000 \times \frac{0.01}{(1 + 0.01)^{60} - 1} \approx \$122.45.$$

Income Annuities

An income annuity provides a sequence of regular equal payments to an individual, the **annuitant**, or to several individuals, the annuitants. The payments may be made over a fixed period of time, in which case the annuity is called an **annuity certain**, or the payments may continue indefinitely until some contingency event occurs. The most common contingency that ends an annuity is the death of the annuitant. An annuity that is terminated by a death is called a **life annuity**; it is good for the lifetime of the annuitant. Therefore, the life expectancy of the annuitant is a major factor in determining the cost of the annuity. Retirement benefits are life annuities and are often variable with a cost-of-living adjustment. By contrast, the cost of an annuity certain can be readily computed without reference to the age or health status of the annuitant.

> **Tidbit**
>
> *Variable annuities are annuities in which the payment size is based on some economic index. This is done to provide protection against inflation.*

EXAMPLE 6.27 The parents of a college student give her a lump sum of $300 with the admonition "This is your pocket money for the term. Make it last." Assuming the term is 10 weeks long, describe how the student can turn this into an income annuity.

Solution

The student could ask her parents to keep the money at home, and send her $30 each week. By giving up the lump sum of $300, the student has effectively purchased an annuity certain with a term of 10 weeks, a payment period of 1 week, and a payment size of $30.

The student in Example 6.27 protected herself from the temptation to spend the pocket money too fast, but she did not earn any interest on the money. Usually, one expects an annuity certain to earn interest. The process involved in an annuity certain is the same as putting a specific amount of money into an interest-paying account and making periodic equal withdrawals, but no deposits, until the account is empty.

Table 6.7 shows the amount that must be paid to an account (that earns a specific interest rate, compounded monthly) in order to receive $1 a month for a specific number of months. The lump-sum payment is called the **present value of the annuity**. The term of the annuity (the number of months payments will be received) determines the row in which to look, and the annual interest rate determines the column. The value in the intersection of the row and column is called the **annuity factor**. For each dollar of monthly income desired, you must pay that many dollars to purchase the annuity. This table assumes that it is an *ordinary annuity;* that is, you will receive your first payment at the *end* of the first month.

TABLE 6.7
DEPOSIT REQUIRED FOR RETURN OF $1 A MONTH FOR *n* MONTHS AT EACH RATE.

Months	3%	4%	6%	8%	10%	12%
12	11.8073	11.7440	11.6189	11.4958	11.3745	11.2551
24	23.2660	23.0283	22.5629	22.1105	21.6709	21.2434
36	34.3865	33.8708	32.8710	31.9118	30.9912	30.1075
48	45.1787	44.2888	42.5803	40.9619	39.4282	37.9740
60	55.6524	54.2991	51.7256	49.3184	47.0654	44.9550
120	103.5618	98.7702	90.0735	82.4215	75.6712	69.7005
180	144.8055	135.1921	118.5035	104.6406	93.0574	83.3217
240	180.3109	165.0219	139.5808	119.5543	103.6246	90.8194
300	210.8765	189.4525	155.2069	129.5645	110.0472	94.9466
360	237.1894	209.4612	166.7916	136.2835	113.9508	97.2183

As an example, suppose you want to purchase an income annuity that provides $1500 a month for 10 years, and the offered rate is 6%. Since the table is based on months, you need to change 10 years to 120 months and look for the annuity factor corresponding to 120 months and 6%. The annuity factor is 90.0735. This means you will need to make a payment of $135,110.25 ($1500 \times 90.0735$). This will provide you with payments of $1500 a month for 10 years, a total of $180,000.

EXAMPLE 6.28 What is the price to purchase a 20-year income annuity paying $500 a month if the return on investment is 8%?

Solution

Consulting Table 6.7, we find the annuity in the 20-year row (240 months) and 8% column is 119.5543. The cost of the annuity would be

$$500 \times 119.5543 = \$59{,}777.72.$$

Over the life of the annuity, the annuitant will receive $120,000.

INITIAL PROBLEM SOLUTION

In 18 years you would like to have $50,000 saved for your child's college education. At 6% interest, compounded monthly, what monthly deposit must be made to accomplish this goal?

Solution

We must determine the payment size for a sinking fund with a future value of $F = \$50{,}000$. The term of the sinking fund is 18 years and the payment period is one month, so the number of payments is

$$n = 12 \times 18 = 216.$$

The interest received per period is

$$i = 0.06 \div 12 = 0.005.$$

It is not specified whether payment is to be made at the beginning of each period or at the end of each period, so we will do it both ways. Plugging the given information into the two formulas, we find

$$PMT = \$128.44$$

if payment is made at the beginning of each month (an annuity due) and

$$PMT = \$129.08$$

if payment is made at the end of each month (an ordinary annuity). In actual practice you would most likely round up to $130 for convenience.

Problem Set 6.5

1. The Russell family decides to save for a vacation by putting aside $300 at the start of each month for the next six months in an account paying 6% interest compounded monthly. When they withdraw the money and interest six months later, how much will they have for the vacation?
2. Theo Amundsen wants to save some money for a big screen television. He puts aside $200 at the start of each month in an account paying 8% interest compounded monthly. After 12 months he will withdraw the money and the interest. How much will he have for the television?
3. Gerry needs to save some money for orthodontia. At the start of each month he puts $150 in an account paying 8% interest compounded monthly. After 10 months how much money will he have saved, including interest?
4. Shawna decides to save money for a down payment on a new car. She sets aside $175 at the start of each month in an account paying 5% interest compounded monthly. After 12 months how much money will she have saved, including interest?
5. Find the future value of an ordinary annuity with a term of 25 years, payment period of one month, payment size of $50, and annual interest rate of 6% compounded monthly.
6. Find the future value of an ordinary annuity with a term of 20 years, payment period of one month, payment size of $75, and annual interest rate of 5%.
7. Traci makes a deposit of $60 at the end of each month into an account paying 5% interest. How much will she have in the account at the end of two years?
8. Brian saves for a sailboat by making payments of $100 to an account at the end of each month. If the account pays 6% interest, how much will be in the account at the end of five years?
9. Find the future value of an annuity due with a term of 25 years, payment period of one month, payment size of $50, and an annual interest rate of 6%.
10. Find the future value of an ordinary annuity with a 20-year term, payment period of one month, payment size of $80, and an annual interest rate of 7%.
11. If an ordinary annuity has a five-year term, a payment period of one month, $80 payments, and an annual interest rate of 6%, what is its future value?
12. What is the future value of an annuity due with a payment of $75 a month, a term of five years, and an annual interest rate of 5.4% compounded monthly?
13. Find the future value of an annuity due with a term of 25 years, payment period of three months, payment size of $100, and an annual interest rate of 6%.
14. Find the future value of an ordinary annuity with a 20-year term, payment period of three months, payment size of $250, and an annual interest rate of 7%.
15. If an ordinary annuity has a 10-year term, a payment period of six months, $200 payments, and an annual interest rate of 6%, what is its future value?
16. What is the future value of an annuity due that has a payment of $300 every four months, a term of 10 years, and an annual interest rate of 5.4%?
17. What monthly payment is required for an annuity due to accumulate $25,000 in a sinking fund in five years with 12% annual interest?
18. If an ordinary annuity is used, what size monthly payment is required if $6,000 is to be accumulated in a sinking fund in four years with 10% annual interest?
19. Suppose you decide to use a sinking fund to save $10,000 for a car. If you plan on 60 monthly payments and you receive 1% interest per month, what is the required payment for an annuity due?
20. Suppose you use a sinking fund to save $12,000 for a car. If you plan on 60 monthly payments and you receive 0.5% interest per month, what is the required payment for an ordinary annuity?

21. A family decides to use a sinking fund to save $15,000 for the down payment on a house. If they plan on 48 monthly payments and receive 0.8% interest per month, what is the required payment for an ordinary annuity?
22. A small company sets up a sinking fund to save $50,000 for a new phone system. If the company makes 60 monthly payments and receives 0.75% interest per month, what is the required payment for an annuity due?
23. The county government sells a series of bonds that have a face value of $8,000,000 that are to be redeemed in 10 years. The county sets up a sinking fund with 6% APR to cover the bonds. If monthly payments begin with the current month (an annuity due) what minimum monthly payment is required so that the bonds will be covered by the date of redemption?
24. The school district issues $6.5 million in bonds for new high school computer laboratories. The district will issue the bonds on the same date as the beginning of the district's budget year and start a sinking fund to redeem the bonds when they mature. Ten annual payments will be made beginning next year (an ordinary annuity). What size payment is required if the district can get 9% interest?
25. What is the price to purchase a 20-year income annuity paying $500 per month if the return on investment is 10%?
26. What is the price to purchase a 20-year income annuity paying $1000 per month if the return on investment is 6%?
27. What amount is needed to purchase a 30-year income annuity paying $1500 per month if the return on investment is 8%?
28. Find the purchase price of a 15-year income annuity paying $1250 per month if the return on investment is 6%.
29. The Jordans set up a fund for their son's college expenses. They will put $40 a month into the fund beginning on his third birthday and ending on his 18th birthday. How much will be in the fund when he turns 18 if the fund earns 5.4% annually, compounded monthly? (The first payment is on his third birthday and the last is on his 18th birthday.)
30. A couple decides to save money for the down payment on a home by putting $250 into an account at the beginning of each month for the next four years. How much will they have available if the annual interest rate is 6%?

Extended Problems

31. The Browns establish an account that they will use to pay cash when buying a new car. They put $5000 into the account when they open it and will deposit $200 at the end of each month for the next three years. If their account has an annual interest rate of 5.75%, how much will they have in the fund at the end of three years? (*Hint:* Treat this as *two* problems.)
32. Suppose you want to have $10,000 available to buy a new car in three years and decide to use the following strategy: You will make a lump sum deposit into an account now and then make monthly payments at the beginning of each month. Each of the sources of funds should generate $5000 toward the total. If you can get a 6% rate of interest, how much should you deposit as a lump sum, and what should the monthly payment be?
33. On her 35th birthday, Merrie decides that when she is 50 she would like to take two years off from work so that she can travel and see the world. To be able to do this, she will make monthly deposits into a fund that will pay an annual rate of 6%, and then when she turns 50 start making monthly withdrawals. Merrie estimates that she will need $2500 per month for her expenses when she starts to travel.
    (a) How much will Merrie need to have available in the account when she is 50?
    (b) How much should Merrie deposit at the beginning of each month to achieve her goal?
34. When Ted begins the seventh grade, his mother begins saving for his college living expenses. Her plan is to make monthly deposits into an account earning 6% interest until Ted begins college in six years; then she will make regular withdrawals from the account to cover his living expenses. She estimates that Ted will need $600 a month for four years, including summer. How much should she deposit each month?
35. Research the options open for individuals who want to establish tax-sheltered annuities. What are eligibility requirements? What are the limitations on the amounts that can be deposited? What other restrictions exist?
36. What are the advantages of tax-sheltered annuities to employees? To employers? To the government?
37. Why would a Swiss annuity (bought in Switzerland and paying Swiss francs) be preferable to an annuity purchased in the United States?
38. One of the difficulties with long-term investing is that the entire nature of the economy can change over the course of a lifetime. Early in this century stock in railroad companies was considered by experts to be the soundest of investments. Research the economic fortunes and stock prices of the Pennsylvania Railroad (now Penn Central) during this century.

✓Chapter 6 Problem

You are buying a car for $10,000 from a dealer who will give you a loan for five years at 10% interest. As an extra incentive he offers either $1200 cash back on the 10% loan or a 6% interest rate (which is a rate that you could expect to get if you deposited funds in a bank or money market account). Which deal should you choose?

SOLUTION

One approach to making a rational decision is to compare the total amount you would pay under each option. That means you have to first find the amount of the monthly payments for each choice. From this information, you can compute the total for *all* payments.

Since the interest rate is a whole percent, we can use a table rather than computing the monthly payment from a formula. Table 6.5 can be used for this purpose. The part of the table we are interested in is the following:

Amount per $1000	
Percent	*5 Years*
6	19.332802
10	21.247045

We see that if the interest were 10% on a five-year loan, then the monthly payments would be $21.247045 for a $1000 loan. Multiplying by 10 gives payments of $212.48 for a $10,000 loan. (Note that we rounded up.) The total of the 60 monthly payments is $60 \times 212.48 =$ $12.748.80.

If the interest rate were 6%, then the monthly payments would be $19.332802 for a $1000 loan, which turns into $193.34 for a $10,000 loan. The total of the 60 monthly payments is $60 \times 193.34 =$ $11,600.40. (Note that the difference between the monthly payments is $19.14.)

One way to look at the problem is that you have a choice either to have $1200 now or have an extra $19.14 a month for the next 60 months. The total amount of money that you will keep by going with the lower interest is $19.14 $\times 60 =$ $1148. Thus you should take the $1200 rather than the lower interest rate since you will have $1200 to spend now or to invest during the entire 60 months. If you can invest the $1200 at 6% compounded monthly for the next five years, it would be worth $1618. The net amount you would pay for the car would be $60 \times 212.48 - 1618 =$ $11,130.80 compared to the $11,600.40 you would pay for the car with the lower rate.

✓Chapter 6 Review

Key Ideas and Questions

The following questions review the main ideas of this chapter. Write your answers to the questions, and refer to the pages listed to make certain that you have mastered these ideas.

1. How does simple interest differ from compound interest? 293–296

   Which seems to be the more fair method of paying interest? 293–296

   How can you compare different compound interest programs? For example, how would you compare borrowing money at 8% compounded annually with 7.6% compounded monthly? 298–299

2. Why is add-on interest less advantageous to the customer than simple interest? 304–305

   How can you compare different loan arrangements such as add-on interest and rent-to-own at different rates? 304–305

3. Describe the history of an amortized loan and explain why the amount of principal paid increases each month. 309–310

4. Describe the process of buying a house and the affordability guidelines designed to prevent a buyer from becoming overextended. 316–320

5. How does the purchase of an income annuity work? 327–328

   Why is an annuity like an amortized loan in reverse? 325–326

Vocabulary

Following is a list of key vocabulary for this chapter. Mentally review each of these items, write down the meaning of each term, and use it in a sentence. Then refer to the pages listed by number, and restudy any material that you are unsure of before solving the Chapter Six Review Problems.

Section 6.1

Simple Interest 293
Principal 293
Interest Rate 293
Time Period 293
Future Value for Simple Interest 294
Ordinary Interest 295
Compound Interest 296

Section 6.2

Section 6.3

Section 6.4

Section 6.5

Chapter 6 Review Problems

1. Suppose that the principal on a loan is \$650, the interest rate is 8%, and the loan is made for a period of four years. What is the interest on the loan if it is calculated as simple interest?

2. Suppose that the terms of the loan in Problem 1 are changed so that the rate is still 8%, but the loan was made June 1, 1997 and was to be repaid September 1, 1999. How much interest should be charged if it is calculated as ordinary interest?

3. Suppose a loan of \$650 is made and is to be repaid in four years with an interest rate of 8% compounded yearly. What total amount will be due in four years? How much of this is interest?

4. Suppose that the interest in Problem 3 is to be compounded monthly. What total amount will be due in four years? How much of this is interest? Compare this with the interest payments in Problem 1 and Problem 3.

5. What is the Effective Annual Rate of an investment at 9% if interest is compounded monthly?

6. Suppose that during a 30-day month the average daily balance on a credit card is \$412, and the interest rate is 19%. What is the finance charge for the month if the billing period is 30 days?

7. Suppose that on June 1 the balance on a credit card is \$0. On June 20 the card is used to purchase a television set for \$612. What will be the average daily balance for the month of June if this is the only transaction? What will the finance charge be if the interest rate is 21%?

8. What is the monthly payment if you borrow \$750 with an interest rate of 15% for 36 months, and the interest is computed as add-on interest? What is the approximate APR for this loan?

9. A television set and a VCR are offered for \$46 per month on a rent-to-own agreement. After 24 months, the television and VCR are yours. Suppose that the actual cost of the system is \$819. What is the approximate APR that you have paid?

10. A loan of \$61,000 is made at an interest rate of 9% and is to be paid over a 10-year period. The monthly payments are \$772.72 per month. Chart the first three months' history of this amortized loan.

11. Use a table to compute the monthly payment on a 10-year loan of \$16,000 at 11% interest per year.

12. Using the formula, find the monthly payment on a three-year loan of \$5125 at 10% interest.

13. Suppose that Pat and Chris are hoping to buy a house. Their combined income is \$51,000 per year. Using the affordability guidelines given in the text, what is the maximum purchase price they should pay for a home?

14. In Problem 13, what are the high and low estimates of what they can afford to pay for a monthly payment, including mortgage, taxes, and insurance?

15. Pat and Chris have found a house they like for \$110,000. They have \$14,000 in savings for the down payment and closing costs. The closing costs are expected to be \$3000. The insurance for a year is \$450, and they can assume taxes will be 3.5% of the value of the home. The bank will give them a 30-year loan at 8%. What will the monthly payments be? Can they afford these payments? Refer to Problem 14.

16. Juanita sets up a savings account in which she pays $40 on the first of each month beginning with February 1. The last payment will be made November 1, and the money will be withdrawn on December 1. How much money will she have if the interest rate is 6%?

17. In order to have $50,000 in 10 years, how much should be deposited in an account every six months if interest is compounded semiannually with a rate of 6%? Assume the deposit is made at the beginning of each payment period.

18. Find the future value of an ordinary annuity with a term of 20 years, an annual interest rate of 5%, and a monthly payment size of $60.

19. Using the annuity table, Table 6.7, find the amount of money needed to purchase a 25-year annuity paying $800 per month if the return on investment is 6%.

PROBLEM SET 6.1

1. **(a)** \$126 **(b)** \$240 **(c)** \$864.50
3. **(a)** \$144 **(b)** \$420 **(c)** \$154.38
5. \$33.12
7. \$87.87
9. **(a)** \$84 **(b)** \$178.13
11. **(a)** \$427.70 **(b)** \$1170
13. \$2300
15. \$3975
17. \$3068.54
19. 3 months-\$1545
6 months-\$1591.35
9 months-\$1639.09
1 year-\$1688.26
21. (iii) 4.9 compounded monthly
23. 8.27%
25. 4%
27. 43.75%
29. \$2970.59
31. 8.3%
33. 6 years
35. 11 years
37. \$12,609.67
39. **(a)** \$625,000 **(b)** \$416,667 **(c)** \$250,000
41. Approximately 8.45%
47. 8 years. \$2500 would become \$5122.30 in 8 years.
49. 16 years. After 16 years \$2500 would become \$5129.17.

PROBLEM SET 6.2

1. **(a)** \$2.70 **(b)** \$5.39
3. **(a)** \$3.99 **(b)** \$3.72
5. \$4.27, \$324.77
7. \$2.03, \$149.88
9. \$1.51, \$137.43
11. **(a)** \$120.45, \$1.32
(b) \$181.31
13. **(a)** \$244.06, \$2.99
(b) \$287.94
15. \$36.56
17. \$28.92
19. \$31.77
21. \$3500
23. \$795
25. **(a)** 20.3% **(b)** 11.1%
(c) 18.0% **(d)** 14.7%
27. \$168
29. **(a)** 11.5% **(b)** 10%
31. **(a)** 14.5% **(b)** 16.5%

33. APR is approximately 22.5%.

35. Add-on rate is 14%.
APR is approximately 25.2%.

37. Add-on rate is 16%.
APR is approximately 28.8%.

39. Add-on rate is 10%.
APR is approximately 18%.

41. 38%

PROBLEM SET 6.3

1. $1974.17

3. $12,754.29

5.

Payment	Interest	Net Payment	Balance
$111.23	$50.00	$61.23	$4938.77
$111.23	$49.39	$61.84	$4876.93
$111.23	$48.77	$62.46	$4814.47

7.

Payment	Interest	Net Payment	Balance
$123.78	$6.25	$117.53	$482.47
$123.78	$5.03	$118.75	$363.72
$123.78	$3.79	$119.99	$243.73
$123.78	$2.54	$121.24	$122.49
$123.77	$1.28	$122.49	$0.00

The last payment is $123.77

9. $132.16

11. Write $18,000 as the sum of the largest amounts that can be found in the table, and add the corresponding payments for a total of $382.47.

13. $400.40

15. $189.87

17. $264.30

19. $20.76

21. $242.77

23. $258.18

25. $179.75

27. $120.58

29. $283.27

31. $55,572.47

33. Using a factor of 10.44, we get an approximate answer of $57,500.

35. $12,330

37. 9%

39. 11.5%

45. $41,369.98

PROBLEM SET 6.4

1. The house should cost $105,000 or less.
The total monthly payments should be $729 or less.

3. The house should cost $130,650 or less.
The total monthly payments should be $907 or less.

5. The house should cost $159,000 or less.
The total monthly payments should be $1104 or less.

7. $1387 (high), $913 (low)

9. $1900

11. $2300

13. $115,000

15. $90,000, $30,000

17. $155,000

19. $740.61

21. $792.47

23. $965.16

25. $729

27. $1291.99

29. $1107.06

PROBLEM SET 6.5

1. $1831.76

3. $1556.11

5. $34,649.70

7. $1511.16

9. $34,822.94

11. $5581.60

13. $23,223.51

15. $5374.07

17. $303.08

19. $121.23

21. $257.56

23. $48,573.53 or $48,574

25. $51,812.30

27. $204,425.25

29. $11,145.99

31. $13,776.79

33. **(a)** $56,407.25
(b) $193.96

CHAPTER SIX REVIEW PROBLEMS

1. $208

3. $884.32, $234.32

5. 9.38%

7. $224.40, $3.87

9. about 31%

11. $220.40

13. $153,000

15. mortgage $726.43
taxes and insurance $358.33
total monthly payment $1084.76
Yes, the low estimate of what they can afford is $1062; the high estimate is $1615.

17. $1806.59

19. $124,165.52